Lecture Notes in Mathematics

Edited by A. Dold and B. Eckmann

668

The Structure of Attractors in Dynamical Systems

Proceedings, North Dakota State University,
June 20–24, 1977

Edited by
N. G. Markley, J. C. Martin and W. Perrizo

Springer-Verlag
Berlin Heidelberg New York 1978

Editors
Nelson G. Markley
Department of Mathematics
University of Maryland
College Park, MD 20742/USA

John C. Martin
William Perrizo

Department of Mathematics
North Dakota State University
Fargo, North Dakota 58102/USA

Library of Congress Cataloging in Publication Data
Main entry under title:

The Structure of attractors in dynamical systems.

(Lecture notes in mathematics ; 668)
Bibliography: p.
Includes index.
1. Differentiable dynamical systems--Congresses.
2. Differential equations--Congresses. 3. Ergodic theory--Congresses. 4. Measure theory--Congresses.
I. Martin, John Calhoun, 1945- II. Markley, Nelson Groh, 1940- III. Perrizo, W., 1943- IV. Series: Lecture notes in mathematics (Berlin) ; 668.
QA3.L28 no. 668 [QA614.8] 510'.8s [516'.36] 78-13670

AMS Subject Classifications (1970): 14K05, 25A65, 28A50, 34A25, 34C05, 34C35, 35L65, 47A35, 54H20, 57D30, 57D45, 57D65, 57D70, 58A25, 58C15, 58C25, 58F05, 58F15, 58F20, 58F99, 73H05, 82A25, 86A25

ISBN 3-540-08925-X Springer-Verlag Berlin Heidelberg New York
ISBN 0-387-08925-X Springer-Verlag New York Heidelberg Berlin

Printing and binding: Beltz Offsetdruck, Hemsbach/Bergstr.
2141/3140-543210

PREFACE

This volume contains papers in dynamical systems and ergodic theory by the participants in the Regional Conference held at North Dakota State University during the week of June 20-24, 1977. During the conference, Professor Rufus Bowen presented ten lectures on "The Structure of Attractors in Dynamical Systems," which will appear in print elsewhere. We are grateful to the National Science Foundation for their support of the conference, which brought together many of the recent contributors to this theory; we hope that this Springer Lecture Notes Volume will provide still further impetus to research in this rapidly growing area.

We wish to express our gratitude to Rufus Bowen, whose outstanding lectures provided a focal point for the conference and were largely responsible for its success. Thanks are due to Pat Berg for her fine typing job.

Nelson Markley
John Martin
William Perrizo

November 28, 1977
Fargo, North Dakota

This volume

is dedicated to the memory of

RUFUS BOWEN

January 15, 1945 - July 30, 1978

TABLE OF CONTENTS

FINITISTIC CODING FOR SHIFTS OF FINITE TYPE

Roy Adler and Brian Marcus

§0. Introduction

Suppose that $\sigma\colon \Sigma_A \to \Sigma_A$ and $\sigma\colon \Sigma_B \to \Sigma_B$ are mixing shifts of finite type with the same entropy. This means that the matrices A and B are aperiodic (i.e., A^n, B^m, are strictly positive for some m, n) and their largest eigenvalues, $\lambda(A)$ and $\lambda(B)$, are equal. (Their common value is a positive number by virtue of the Perron-Frobenius theorem.) See [8], [5] for more information.

What can be said about the relationship between Σ_A and Σ_B? Well, first of all, they need not be topologically conjugate. For example $A = \begin{pmatrix} 1 & 1 \\ 1 & 1 \end{pmatrix}$ and $B = \begin{pmatrix} 0 & 1 & 1 \\ 1 & 0 & 1 \\ 1 & 1 & 0 \end{pmatrix}$ are both aperiodic and have entropy $\log 2$ (since they each have row sum 2); however, they are not topologically conjugate because Σ_A has a fixed point but Σ_B does not (look at the diagonals). (R. Williams ([12]) has studied the problem of classifying shifts of finite type under topological conjugacy.)

Nevertheless, it is now well known that, from the measure-theoretic point of view, something quite strong can be said. By [9], there are unique measures μ_A and μ_B on Σ_A and Σ_B (resp.) which have maximal entropy (e.g., the measure theoretic entropy $h(\mu_A)$ of $\sigma\colon \Sigma_A \to \Sigma_A$ equals the topological entropy of $\sigma\colon \Sigma_A \to \Sigma_A$). With respect to these measures, the shifts are well-known to be Bernoulli and so by Ornstein's isomorphism theory they are measure-theoretically conjugate ([7]). (This is weaker than a topological conjugacy.) One way of looking at what our main theorem says is: the conjugacy (guaranteed by Ornstein's theory) can be chosen to be a homeomorphism a.e. and in particular gives an independent proof of this special case of

Ornstein's theorem.

<u>Definition</u>: If $\varphi: X \to X$ and $\psi: Y \to Y$ are homeomorphisms of compact metric spaces with unique measures of maximal entropy, one says that φ is an <u>almost conjugate extension</u> of ψ (or ψ is an <u>almost conjugate factor</u> of φ) if there is a map $\pi: X \to Y$ such that $\pi \circ \varphi = \psi \circ \pi$, π is continuous, onto, boundedly finite-to-one, and almost everywhere 1 - 1. So, π is close to being a topological conjugacy, without actually being one.

Main Theorem: <u>If Σ_A and Σ_B are aperiodic and have the same entropy, then Σ_A and Σ_B have a common almost conjugate extension Σ_C which is also an aperiodic shift of finite type</u>.

We abbreviate the conclusion of the theorem by the notation: $\Sigma_A \sim \Sigma_B$. This is an equivalence relation. The details of the proof of the main theorem will appear elsewhere.

Pictorially the main theorem says:

$$\begin{array}{ccc} & (\sigma, \Sigma_C) & \\ {}^{\pi_1}\swarrow & & \searrow^{\pi_2} \\ (\sigma, \Sigma_A) & & (\sigma, \Sigma_B) \end{array}$$

Observe that $\pi_1 \circ \pi_2^{-1}$ is a homeomorphic a.e. which conjugates $\sigma: \Sigma_B \to \Sigma_B$ with $\sigma: \Sigma_A \to \Sigma_A$. It sends the measure μ_B to a measure on Σ_A with the same entropy -- but this measure must be μ_A by uniqueness of the measure of maximal entropy. Thus, $\pi_1 \circ \pi_2^{-1}$ is a measure-preserving homeomorphism a.e.

Since $\pi_1 \circ \pi_2^{-1}$ is a homeomorphism a.e., one sees that Σ_B can be coded to Σ_A finitistically in the following sense: for $b \in \Sigma_B$ but outside a set of measure zero, every finite block of $\pi_1 \circ \pi_2^{-1}(b)$ may be determined by knowledge of only a finite (though not uniformly finite) block of b. (See [11].)

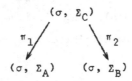

As a corollary of the main theorem, we get (see [8] for the notation of period) a "$\sim$" classification for shifts of finite type which are merely irreducible.

Corollary: <u>If Σ_A and Σ_B are irreducible and have the same entropy and period, then Σ_A and Σ_B have a common almost conjugate extension Σ_C which is also an irreducible shift of finite type.</u>

Partial results toward our main theorem were obtained earlier. Adler and Weiss [2] proved the main theorem in the case that Σ_A and Σ_B represent Markov partitions for hyperbolic total automorphism of T^2. Adler, Goodwyn, and Weiss ([1]) proved the main theorem in the case that the entropy is the log of a positive integer. Parry ([10]) proved the theorem except for the $1-1$ a.e. condition.

Bowen's work on Markov partitions ([3], [4], [5]) shows that every mixing Axiom A basic set (e.g., hyperbolic toral automorphisms, connected Axiom A attractors) is an almost conjugate factor of a mixing shift of finite type. Thus, as an application of the main theorem, one sees that two mixing Axiom A basic sets with the same entropy are equivalent in our sense. In particular they are conjugate by a measure-preserving homeomorphism a.e. (where the measures involved are those of maximal entropy).

The proof of the main theorem has two major steps:

<u>Step 1</u>: The theorem is true if both A and B have fixed states (i.e., both has a 1 on the diagonal).

<u>Step 2</u>: Given an aperiodic A, there exists an aperiodic M such that $\Sigma_A \sim \Sigma_M$ and M has a fixed state. (<u>Note</u>: necessarily Σ_A and Σ_M must have the same entropy; also irreducibility of A would not be sufficient.)

Of course, Steps 1 and 2 and the transitivity of the relation "$\sim$" yield the main theorem. We actually believe that the method of step 1 works without the fixed state assumption, but we are not able to do without it. At any rate, the fixed state assumption does make the problem simpler, although we shall not fully indicate why in this paper. We will talk mainly about step 2 because it involves ideas about Markov partitions and attractors -- the topic of this conference.

§1. Outline of Step 1

The proof rests on

Furstenberg's Lemma ([10]): <u>If</u> Σ_A <u>and</u> Σ_B <u>are irreducible, then they have the same entropy if and only if there exists a positive integral matrix</u> F <u>such that</u>

$$FA = BF.$$

This does not require the fixed state assumption. Now, the matrix F tells us how to construct the alphabet $\mathcal{C}$ for Σ_C.

$$\mathcal{C} = \{(I,J,K): \ I \in \mathcal{A}, \ \ J \in \mathcal{B}, \ \ 1 \le K \le F_{I,J}\}$$

where $\mathcal{A}$, $\mathcal{B}$ are the alphabets for Σ_A, Σ_B (resp.). Now, Furstenberg's lemma implies that one can define transition rules on $\mathcal{C}$ so that

0. if $(I,J,K) \to (I',J',K')$ then $I \to I'$ and $J \to J'$,
1. if $(I,J,K) \in \mathcal{C}$ and $J \to J'$ then $\exists!\ I'$, K' such that $(I,J,K) \to (I',J',K')$ and simultaneously
2. if $(I',J',K') \in \mathcal{C}$ and $I \to I'$, then $\exists!\ J$, K such that $(I,J,K) \to (I',J',K')$.

Let C denote the matrix which represents the transition rules. Then there are natural projection maps $\pi_1: \Sigma_C \to \Sigma_A$ and $\pi_2: \Sigma_C \to \Sigma_B$. These maps are continuous, and they are onto and boundedly finite-to-one by 0, 1 and 2 above. (See [1] for a little more detail.)

However, C may not be aperiodic and π_1 and π_2 may not be $1-1$ a.e. In order to get these things, one has to impose more requirements on the matrix C, consistent with 0, 1 and 2 above. For this, one uses the notion of resolving words, which is discussed in [1]. Namely, letting I^* and J^* be fixed states of A and B resp., we rig the transition rules (i.e., the matrix C) in such a way that there exists an n such that $I^*I^*\ldots I^*$ and $J^*J^*\ldots J^*$ are each resolving words of length n for π_1 and π_2 resp. More specifically this means that

3. if $(I_1,J^*,K_1)(I_2,J^*,K_2)\ldots(I_n,J^*,K_n)$ is an allowable C-word then $I_n = I^*$ and $K_n = 1$ and
4. if $(I^*,J_1,K_1)(I^*,J_2,K_2)\ldots(I^*,J_n,K_n)$ is an allowable C-word then $J_1 = J^*$ and $K_1 = 1$.

The fixed states are useful because you can get to a fixed state and stay there. One can see that 0, 1, 2, 3, 4 give the $1-1$ a.e. and aperiodicity conditions.

§2. Outline of Step 2

What we actually show is that given A, there exists an M with a fixed state such that Σ_A and Σ_M have a common almost conjugate <u>factor</u> (rather than an extension). This factor turns out to be sufficiently nice so that inside $\{(\underline{x},\underline{y}) \in \Sigma_A \times \Sigma_M\colon \rho_1(\underline{x}) = \rho_2(\underline{y})\}$ (where ρ_1 and ρ_2 are the factor maps) one can find an aperiodic shift of finite type Σ_C and the natural projection maps $\pi_1\colon \Sigma_C \to \Sigma_A$, $\pi_2\colon \Sigma_C \to \Sigma_B$ will be almost conjugate maps. So $\Sigma_A \sim \Sigma_M$.

In the special case that Σ_A represents (the transition rules for) a Markov partition of a hyperbolic toral automorphism $\psi\colon T^n \to T^n$, the proof is particularly simple:

As mentioned before, by Bowen's work ([5]) $\psi\colon T^n \to T^n$ is an almost conjugate factor of $\sigma\colon \Sigma_A \to \Sigma_A$. So, it suffices to find another Markov partition $\mathcal{M}$ for ψ which has a ψ-fixed point (e.g., 0) in int $\mathcal{M}$ (i.e., the interior of some element of $\mathcal{M}$). Well, let $\mathcal{A}$ be the original Markov partition. Since the periodic points are dense, there is a periodic point p such that $0 \in \text{int}\,(\mathcal{A}+p)$. (The "+" sign is the group addition on the torus.) Let

$$\mathcal{M} = (\mathcal{A}+p) \vee (\psi(\mathcal{A}+p)) \vee \ldots \vee (\psi^{\ell-1}(\mathcal{A}+p))$$

(where ℓ is the period of p), i.e.,

$$\mathcal{M} = \{\overline{(\text{int } A_{i_0} + p) \cap (\psi(\text{int } A_{i_1} + p)) \cap \ldots \cap (\psi^{\ell-1}(\text{int } A_{i_{\ell-1}} + p))}\}.$$

One can show that $\mathcal{M}$ is indeed a Markov partition and $\partial\mathcal{M} \subset \bigcup_{i=0}^{\ell-1} \psi^i(\partial\mathcal{A}+p)$ where ∂ denotes the union of the boundaries of the elements of the partition. We claim that $0 \notin \partial\mathcal{M}$ whence $0 \in \text{int}\,\mathcal{M}$. Well, if $0 \in \partial\mathcal{M}$ then $0 \in \partial\mathcal{A}+p$; contrary to the choice of p. □

Now we illustrate the general case with an example. What we do is to replace the hyperbolic toral automorphism above by another object.

Let

$$A = \begin{pmatrix} 0 & 1 & 0 \\ 0 & 0 & 1 \\ 1 & 1 & 0 \end{pmatrix}.$$

Its transition rules are

$$1 \to 2$$
$$2 \to 3$$
$$3 \to 2\ 1.$$

Let K be the wedge of 3 circles, (C_1, C_2, and C_3) each representing a state of the matrix A:

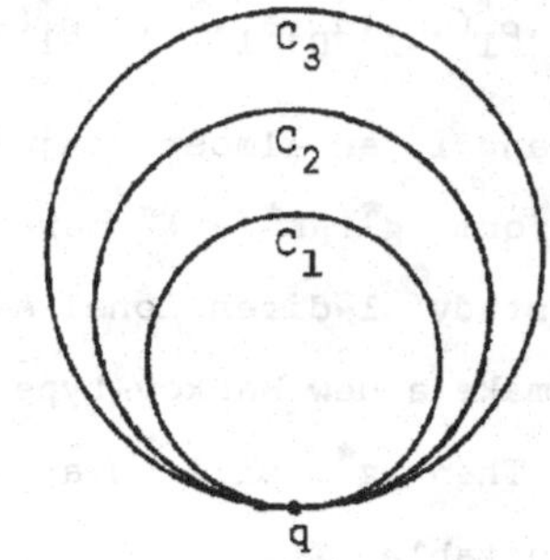

Define $g: K \to K$ to be a continuous (but not 1-1) map such that $g(q) = q$, g reverses orientation and g wraps C_1 once around C_2, g wraps C_2 once around C_3 and g wraps C_3 around C_2 and then C_1. The point is that the decomposition of K into these circles represents a "Markov partition" for g. Thus, in the usual way one gets a map

$$\rho_1^+: \Sigma_A^+ \to K$$

where Σ_A^+ is the one-sided shift of finite type, $\rho_1^+ \circ \sigma = g \circ \rho_1^+$, and ρ_1^+ is an almost conjugate factor map (namely, $\rho_1^+(a_0, a_1, a_2, \ldots) = \bigcap_{n\geq 0} \overline{(\bigcap_{i=0}^n g^{-i}(C_{a_i} - \{q\}))}$ (see [6]). One can extend ρ_1^+ to inverse limits in order to get an almost conjugate factor $g^*: K^* \to K^*$ of $\sigma: \Sigma_A \to \Sigma_A$: Recall that the inverse limit

$$K \overset{g}{\longleftarrow} K \overset{g}{\longleftarrow} K \overset{g}{\longleftarrow} \ldots$$

takes place on

$$K^* = \{(x_0, x_1, x_2, \ldots) \in K \times K \times K \times \ldots \quad \text{s.t.} \quad x_{i-1} = g(x_i)\}$$

and the map is $g^*: K^* \to K^*$

$$g^*(x_0, x_1, x_2, \ldots) = (g(x_0), x_0, x_1, x_2, \ldots).$$

Then the map $\rho_1: \Sigma_A \to K^*$

$$\rho_1((a_i)) = (\rho_1^+(a_0,a_1,\ldots),\rho_1^+(a_{-1},a_0,a_1,\ldots),\rho_1^+(a_{-2},a_{-1},a_0,a_1,\ldots),\ldots)$$

satisfies $\rho_1 \circ \sigma = g^* \circ \rho_1$ and is an almost conjugate factor map. The dynamical systems of the form $g^*: K^* \to K^*$ are precisely the models that R. Williams uses to study 1-dimensional Axiom A attractors ([13]). Now, we want to make a new Markov-type partition for g^* which has a fixed state. Then g^* will be a common almost conjugate factor of Σ_A and some suitable Σ_M.

Chop each circle into 3 equal pieces as indicated:

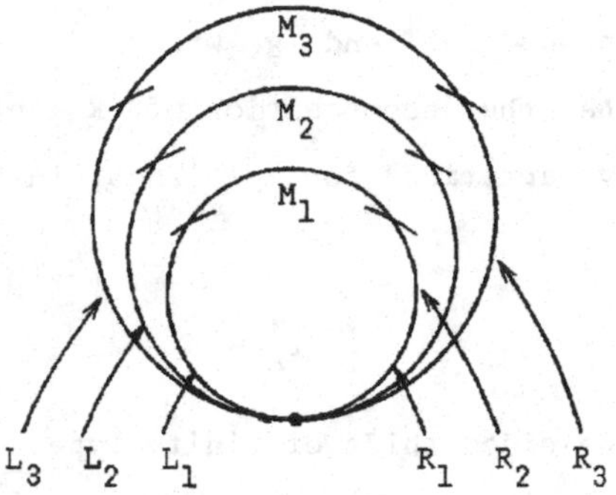

We make the following new alphabet

$$\mathcal{B} = \{M_i,\ i = 1,2,3\} \cup \{L_i \cup R_j,\ i,j = 1,2,3\}.$$

So $\mathcal{B}$ consists of the "middle" states and some "two-pronged" states. Of course, some of these states overlap in their interiors but that will work itself out later. Now, there is a natural set of transition rules on the alphabet $\mathcal{B}$ such that $\forall\ I \in \mathcal{B}$

$$g(I) = \bigcup_{I \to I'} I'$$

and

$$\text{if } I \to I' \text{ and } I'', \text{ then } I' \cap I'' \subset \partial I' \cap \partial I''.$$

Namely,

(1) $$M_1 \to M_2$$

(2) $$M_2 \to M_3$$

(3) $$M_3 \to R_2 \cup L_1$$

and each two pronged state goes to a two-pronged state and possibly M_1 or M_2. For example,

(4) $$R_2 \cup L_1 \to L_3 \cup R_2$$

(5) $$L_3 \cup R_2 \to R_1 \cup L_3,\ M_1$$

(6) $$R_1 \cup L_3 \to L_2 \cup R_1,\ M_1$$

(7) $$L_2 \cup R_1 \to R_3 \cup L_2$$

(8) $$R_3 \cup L_2 \to L_2 \cup R_3,\ M_2.$$

Note that $R_3 \cup L_2$ is a fixed state. What we have done is to use the 2-cycle $2 \to 3 \to 2$ in A to get a fixed state by creating a state $(R_3 \cup L_2)$ which is composed of part of C_2 and part of C_3.

Unfortunately, the alphabet $\mathcal{B}$ together with its natural transition rules is too big -- it's not irreducible. We let $\mathcal{M}$ be the irreducible component which contains the fixed state $R_3 \cup L_2$. It is easily checked that $\mathcal{M}$ consists of the 8 states enumerated above and hence covers K. Its transition matrix

$$M = \begin{pmatrix} 0 & 1 & 0 & 0 & 0 & 0 & 0 & 0 \\ 0 & 0 & 1 & 0 & 0 & 0 & 0 & 0 \\ 0 & 0 & 0 & 1 & 0 & 0 & 0 & 0 \\ 0 & 0 & 0 & 0 & 1 & 0 & 0 & 0 \\ 1 & 0 & 0 & 0 & 0 & 1 & 0 & 0 \\ 1 & 0 & 0 & 0 & 0 & 0 & 1 & 0 \\ 0 & 0 & 0 & 0 & 0 & 0 & 0 & 1 \\ 0 & 1 & 0 & 0 & 0 & 0 & 0 & 1 \end{pmatrix}$$

Now, M is the matrix we want.

One gets in a natural way a map $\rho_2^+ : \Sigma_M^+ \to K$ namely,

$$\rho_2^+(I_0, I_1, I_2, \ldots) = \bigcap_{i \geq 0} g^{-i}(I_i)$$

$\rho_2^+ \circ \sigma = g \circ \rho_2^+$ and ρ_2^+ is continuous, onto, and boundedly finite-to-one. The latter follows (partly) from the

<u>FACT</u>: If $y \in I \in \mathcal{M}$ and $\forall\, n \geq 0$, $g^n(y) \notin \partial\mathcal{M}$ then $\exists!\ \underline{I} \in \Sigma_M^+$ such that $\rho_2^+(\underline{I}) = y$ and $I_0 = I$.

However, ρ_2^+ is not $1-1$ a.e. because the states of $\mathcal{M}$ intersect so badly. For example, each point of R_2 has at least two (ρ_2^+)-preimages one of whose first symbol is the state $R_2 \cup L_1$ and another whose first symbol is the state $L_3 \cup R_2$. Fortunately, the natural extension of ρ_2^+ to inverse limits is $1-1$ a.e. (and inherits the other properties from ρ_2^+): Recall that the natural extension is

$$\rho_2 : \Sigma_M \to K$$

$$\rho_2((I_i)) = (\rho_2^+(I_0, I_1, \ldots), \rho_2^+(I_{-1}, I_0, I_1, \ldots), \rho_2^+(I_{-2}, I_{-1}, I_0, I_1, \ldots), \ldots).$$

To see that ρ_2 is $1-1$ a.e., one first uses the ergodicity of g^* to show that for almost all $x = (x_0, x_1, x_2, \ldots) \in K^*$

(i) $(\forall\, n)$ x_n and $g^n(x_0)$ are <u>not</u> in $\partial\mathcal{M}$

and

(ii) for infinitely many $k > 0$, $x_k \in M_1$.

These two things plus the FACT above show that (for each such x) for infinitely many k, x_k has a unique ρ_2^+ - inverse image. So, if $\rho_2((I_i)) = x$, then for infinitely many $k > 0$

$$I_{-k}, I_{-k+1}, \ldots$$

is uniquely determined. Thus, (I_i) is uniquely determined. Thus, $g^*: K^* \to K^*$ is an almost conjugate factor of both Σ_A and Σ_M.

The reader may verify that the matrices M and A have the same characteristic polynomials up to factors of the form x^n and products of cyclotomic polynomials. This is no accident.

REFERENCES

1. R. Adler, W. Goodwyn, B. Weiss, Equivalence of topological Markov chains, to appear in Israel J. Math.
2. R. Adler, B. Weiss, Similarity of automorphisms of the torus, Memoirs of the A.M.S., No. 98(1970).
3. R. Bowen, Markov partitions for Axiom A diffeomorphisms, Amer. J. Math. 92(1970), 725-747.
4. R. Bowen, Markov partitions and minimal sets, Amer. J. Math. 92 (1970), 907-918.
5. R. Bowen, Equilibrium States and the Ergodic Theory of Anosov Diffeomorphisms, Springer-Verlag Lecture Notes #470.
6. M. Denker, C. Grillenberger, K. Sigmund, Ergodic Theory on Compact Spaces, Springer-Verlag Lecture Notes #532.
7. N. Friedman, D. Ornstein, On the isomorphism of weak Bernoulli transformations, Advances in Math. 5(1970), 365-394.
8. E. Gantmacher, Theory of Matrices, Vol. 2, New York, 1964
9. W. Parry, Intrinsic Markov chains, Transactions of the A.M.S. 112 (1964), 55-66.
10. W. Parry, A finitary classification of topological Markov chains and Sofic systems, preprint.
11. B. Weiss, The structure of Bernoulli systems, Proc. Int'l. Cong. Math., 1974, Vancouver, Vol. 2, p. 324.
12. R. Williams, Classification of shifts of finite type, Annals of Math. 98(1973), 120-153. Erratum: ibid. 99(1974), 380-381.
13. R. Williams, Classification of one-dimensional attractors, Proc. Symp. Pure Math., A.M.S., 1970, Vol. 14 Global Analysis.

Roy Adler
IBM WATSON RESEARCH LABORATORY
YORKTOWN HEIGHTS, N.Y.

Brian Marcus
UNIVERSITY OF NORTH CAROLINA

PERIODIC POINTS AND LEFSCHETZ NUMBERS

by

Steve Batterson*

For a diffeomorphism f on a compact manifold let $N_m(f)$ be the cardinality of the set of fixed points of f^m. Given a manifold and a homotopy class of maps on the manifold, we would like to determine the sequences $\{N_m\}_{m=1}^{\infty}$ for which there exists an Axiom A, no-cycle diffeomorphism in the homotopy class with $N_m = N_m(f)$. In [2] Franks presented results relating $\{N_m(f)\}$ to a set of homological invariants. In this paper we interpret these results to provide a necessary condition to the above problem.

For an Axiom A diffeomorphism f let $N_m^E(f)$ be the cardinality of the set of points of even index that are fixed under f^m and let $N_m^0(f)$ be defined analogously for odd index periodic points. $L(g)$ denotes the Lefschetz number of g.

Theorem. _If_ f _satisfies Axiom A and the no-cycle condition and_ f _is homotopic to_ g _then for any_ $m = 2^d r$,

$$2^{d+1} \mid [N_m^E(f) - N_m^0(f) - L(g^m)].$$

The condition that f and g be homotopic can be weakened to the requirement that their homology zeta functions have the same mod 2 reduction. Thus if f is related to the identity in this manner (e.g. f could be any homeomorphism on a sphere or projective space) and also satisfies Axiom A and no-cycles, then $L(g^m)$ can be replaced by the Euler characteristic. Various applications of these results are presented at the end of the paper.

*Research supported in part by Emory University Research Summer Grant.

§1. Preliminaries

In this section we discuss the zeta function and homology zeta function. We begin by defining these invariants. Further details are in [2] and [9].

The zeta function of a diffeomorphism is defined by $\zeta(t) = \exp(\sum_{m=1}^{\infty} \frac{1}{m} N_m t^m)$. For diffeomorphisms satisfying Axiom A and the no-cycle property, the zeta function is a quotient of two polynomials with integer coefficients and constant term 1 ([4], [5]). This result also holds for the restriction of the map to a basic set. The reduced zeta function Z_i of f on a basic set Λ_i is defined to be the mod 2 reduction of the rational function $\zeta(f|\Lambda_i)$.

The homology zeta function of a diffeomorphism is defined by $\eta(t) = \exp(\sum_{m=1}^{\infty} \frac{1}{m} L(f^m)t^m)$ where the Lefschetz number $L(f) = \sum_{i=0}^{\dim M} (-1)^i \mathrm{tr}(f_{*i})$ and f_{*i} denotes the map induced on the ith homology group with real coefficients. The function $\eta(t)$ is a rational function and it can be shown that

$$\eta(t) = \prod_{i=0}^{\dim M} \det(I - f_{*i}t)^{(-1)^{i+1}}.$$

The Lefschetz numbers $L(f^m)$ provide an algebraic counting of the fixed points of f^m while N_m is a geometric counting of these points. Since $\eta(t)$ is a homological invariant of f and $\zeta(t)$ is a topological conjugacy invariant the following theorem of Franks (in [2]) provides an important connection between the algebraic topology of a diffeomorphism and its dynamics.

Theorem (Franks). Suppose $f\colon M \to M$ satisfies Axiom A and the no-cycle property and has basic sets $\Lambda_1,\ldots,\Lambda_\ell$, then the following are equal.

a. $\prod_{i=1}^{\ell} (Z_i)^{(-1)^{u_i}}$ where u_i = fiber dim $E^u(\Lambda_i)$ = index

b. The reduction mod 2 of $\eta(t)$.

This theorem yields several corollaries describing how the dynamics of a map and its basic sets may change under isotopy.

1. A necessary condition for an isotopy to remove a basic set Λ_i of f while not changing the other basic sets is that $Z_i(f) = 1$.

2. A necessary condition for cancelling two basic sets Λ_i and Λ_j under isotopy without changing the others is that $Z_i^{(-1)^{u_i}} \cdot Z_j^{(-1)^{u_j}} = 1$.

3. A necessary condition for replacing a basic set Λ_i of f by a basic set Λ_i' of an isotopic map g with all other basic sets unchanged is that $Z_i(f)^{(-1)^{u_i}} = Z_i(g)^{(-1)^{u_i'}}$.

In this paper we will determine which sequences of N_m's satisfy these algebraic requirements.

§2. Results and Proofs

In [7] Neuberger gives a formula for the $m^{\underline{th}}$ derivative of a composition of two functions on normed linear spaces. We now state a version of this formula for functions of one variable.

Lemma 1.

$$(f\circ g)^{(m)}(t) = \sum_{\substack{i_1 j_1 + \cdots + i_\ell j_\ell = m \\ 1 \le i_1 < \cdots < i_\ell \le m}} \frac{m!\, f^{(q)}(g(t))\,(g^{(i_1)}(t))^{j_1} \cdots (g^{(i_\ell)}(t))^{j_\ell}}{\prod_{k=1}^{\ell} [(i_k!)^{j_k}(j_k)!]}$$

<u>where</u> $q = \sum_{k=1}^{\ell} j_k$.

The following technical lemma determines the relevant portion of the power series coefficients of a rational zeta function in terms of polynomial coefficients.

Lemma 2. Suppose $\exp(\sum_{m=1}^{\infty} \frac{N_m}{m} t^m) = \frac{b_p t^p + \ldots + b_1 t + 1}{a_n t^n + \ldots + a_1 t + 1}$.

Then

$$N_m = \sum_{\substack{i_1 j_1 + \ldots + i_\ell j_\ell = m \\ 1 \le i_1 < \ldots < i_\ell \le m}} \frac{(-1)^q m(q-1)!\,[(a_{i_1})^{j_1} \ldots (a_{i_\ell})^{j_\ell} - (b_{i_1})^{j_1} \ldots (b_{i_\ell})^{j_\ell}]}{\prod_{k=1}^{\ell} (j_k)!}$$

where $q = \sum_{k=1}^{\ell} j_k$.

Proof: If $\sum_{m=1}^{\infty} \frac{N_m t^m}{m} = \log(b_p t^p + \ldots + b_1 t + 1) - \log(a_n t^n + \ldots + a_1 t + 1) =$ $h(t)$. Then $N_m = \frac{h^{(m)}(0)}{(m-1)!}$.

It remains to compute $h^{(m)}(0)$. We shall assume each $b_i = 0$. Without this assumption the proof is similar. Let $g(t) =$ $a_n t^n + \ldots + a_1 t + 1$. Then

$$\begin{aligned} h^{(m)}(t) &= -\frac{d^{m-1}}{dt^{m-1}} [g'(t)(g(t))^{-1}] \\ &= -\sum_{r=0}^{m-1} \binom{m-1}{r} g^{(r+1)}(t) \frac{d^{m-r-1}}{dt^{m-r-1}} ((g(t))^{-1}) \end{aligned}$$

by Leibniz rule. Thus

$$N_m = -\sum_{r=0}^{m-1} \frac{(r+1)}{(m-r-1)!} a_{r+1} \left[\frac{d^{m-r-1}}{dt^{m-r-1}} ((g(t))^{-1})\right]_{t=0}.$$

Letting $f(t) = \frac{1}{t}$ in Lemma 1 and then evaluating at 0 yields

$$N_m = -\sum_{r=0}^{m-1} (r+1) a_{r+1} \sum_{\substack{i_1 j_1 + \ldots + i_\ell j_\ell = m-r-1 \\ 1 \le i_1 < \ldots < i_\ell \le m-r-1}} \frac{(-1)^q q!\,(a_{i_1})^{j_1} \ldots (a_{i_\ell})^{j_\ell}}{\prod_{k=1}^{\ell} (j_k)!}$$

where $q = \sum_{k=1}^{\ell} j_k$.

Note that each term in N_m contains a factor of the form

$(a_{i_1})^{j_1} \ldots (a_{i_\ell})^{j_\ell}$ where $m = \sum_{k=1}^{\ell} i_k j_k$. The lemma is obtained by computing the coefficients of these terms.

Lemma 3. _If_ N_m _is a sequence of integers such that_ $\exp(\sum_{m=1}^{\infty} \frac{N_m}{m} t^m)$ _is a quotient of polynomials with integer coefficients and constant term_ 1, _then the_ mod 2 _reduction of this rational function is_ 1 _if and only if for any positive integer_ $m = 2^d r$, $2^{d+1} | N_m$.

Proof: Suppose $\dfrac{b_p t^p + \ldots + b_1 t + 1}{a_n t^n + \ldots + a_1 t + 1}$ has integer coefficients such that the reduction mod 2 is equal to 1. If N_m is a sequence of integers such that $\exp(\sum_{m=1}^{\infty} N_m \frac{1}{m} t^m)$ is the above rational function then N_m is given by Lemma 2. It remains to show that $(q-1)![a_{i_1}^{j_1} \ldots a_{i_\ell}^{j_\ell} - b_{i_1}^{j_1} \ldots b_{i_\ell}^{j_\ell}]$ contains at least one more factor of 2 than does $\prod_{k=1}^{\ell} (j_k)!$.

For each k between 1 and ℓ there exists an integer s_k such that $b_{i_k} = a_{i_k} + 2s_k$. So $[a_{i_1}^{j_1} \ldots a_{i_\ell}^{j_\ell} - b_{i_1}^{j_1} \ldots b_{i_\ell}^{j_\ell}] =$ $[a_{i_1}^{j_1} \ldots a_{i_\ell}^{j_\ell} - (a_{i_1} + 2s_1)^{j_1} \ldots (a_{i_\ell} + 2s_\ell)^{j_\ell}]$.

After expanding the expression on the right, $a_{i_1}^{j_1} \ldots a_{i_\ell}^{j_\ell}$ drops out. For each remaining term there exists a k such that j_k contains fewer factors of 2. But $(q-1)!$ contains at least as many factors of 2 as the other terms in $(j_i)! \ldots (j_\ell)!$.

To prove the lemma in the other direction suppose the mod 2 reduction of the rational function is not 1. Let m be the smallest positive integer such that $a_m - b_m$ is not even. Then all terms of N_m are divisible by 2^{d+1} except $-[a_m - b_m]m$.

Theorem. _If_ f _satisfies Axiom A and the no-cycle property and_ g _is the map whose homology zeta function has the same_ mod 2 _reduction as that of_ f _then for any_ $m^d r$, $2^{d+1} \mid [N_m^E - N_m^0 - L(g^m)]$.

Proof:

$$\exp[\sum_{m=1}^{\infty} (N_m^E(f) - N_m^0(f) - L(g^m))\frac{1}{m}t^m]$$

$$= \frac{\exp\left(\sum_{m=1}^{\infty} \frac{N_m^E(f)}{m} t^m\right)}{\exp\left(\sum_{m=1}^{\infty} \frac{N_m^0(f)}{m} t^m\right)} \cdot \frac{1}{\exp \sum_{m=1}^{\infty} \left(\frac{L(g^m)}{m} t^m\right)} .$$

The expression on the right is the reciprocal of $\eta(g)$ and by Frank's theorem the expression on the left has mod 2 reduction equal to that of $\eta(f)$.

The condition that $\eta(f)$ and $\eta(g)$ have the same mod 2 reduction is considerably weaker than homotopy. On spheres and projective spaces all diffeomorphisms are related in this manner and on $T^2 = S^1 \times S^1$ there are two possible reductions. If g is the identity then $L(g^m)$ is the Euler characteristic and we have:

Corollary 1. *Suppose f is an Axiom A, no-cycle diffeomorphism whose homology zeta function has the same mod 2 reduction as the identity map. Then for $m = 2^d r$, $2^{d+1} \mid [N_m^E - N_m^0 - \chi(M)]$ where $\chi(M)$ is the Euler characteristic of the manifold.*

Corollary 2. *An Axiom A, no-cycle diffeomorphism on T^2 has one of the following properties:*

I. *For any $m = 2^d r$, $2^{d+1} \mid (N_m^E - N_m^0)$.*

II. *For any $m = 2^d r$, $2^{d+1} \mid (N_m^E - N_m^0 - L_m)$ where $L_m = 0$ if 3 divides m and otherwise $L_m = 3$.*

Let K_m be the cardinality of the set of periodic points whose prime period is m. We will denote by K_m^E the number of points of prime period m whose index is even. The notation for odd index points is K_m^0. The following two corollaries are for diffeomorphisms satisfying the hypothesis of Corollary 1.

Corollary 3. $2^{d+1} \mid (\sum_{s=0}^{d} K^{E}_{2^s} - \sum_{s=0}^{d} K^{0}_{2^s} - x(M))$.

Corollary 4. If r is an odd integer greater than 1 then

$$2^{d+1} \mid (\sum_{s=0}^{d} K^{E}_{2^s r} - \sum_{s=0}^{d} K^{0}_{2^s r}).$$

Corollary 3 follows immediately from Corollary 1.

Proof of Corollary 4: Suppose r is an odd prime number. Then

$$N^{*}_{2^d r} = \sum_{s=0}^{d} K^{*}_{2^s r} + \sum_{s=0}^{d} K^{*}_{2^s} .$$

So $2^{d+1} \mid (\sum_{s=0}^{d} K^{E}_{2^s r} + \sum_{s=0}^{d} K^{E}_{2^s} - \sum_{s=0}^{d} K^{0}_{2^s r} - \sum_{s=0}^{d} K^{0}_{2^s} - x(M))$. Thus

$2^{d+1} \mid (\sum_{s=0}^{d} K^{E}_{2^s r} - \sum_{s=0}^{d} K^{0}_{2^s r})$ by Corollary 3.

The proof of Corollary 4 follows by induction on the number of prime factors of r.

We now state some applications of the corollaries. Each application is for diffeomorphisms satisfying the hypothesis of Corollary 1. These results are related to those in [3].

Application 1. If the Euler characteristic of M is greater than the total number of even index periodic points whose periods are powers of 2, then f has an infinite number of odd index periodic points whose periods are powers of 2.

Application 2. If the Euler characteristic of M is less than the negative of the total number of odd index periodic points whose periods are powers of 2, then f has an infinite number of even index periodic points whose periods are powers of 2.

Application 3. Fix $r > 1$ and odd and suppose for any d, f has no even (odd) index periodic points of period $2^d r$. Then f has either 0 or an infinite number of odd (even) index points whose periods are of the form $2^d r$.

We close by translating into the framework of this paper, Franks' necessary conditions for adding, subtracting, cancelling, and substituting basic sets of Axiom A, no-cycle diffeomorphisms under isotopy. These results may indicate some of the dynamical behavior which occurs during bifurcation. For a basic set Λ_i, let N_m^i be the number of points in Λ_i fixed under f^m.

1. A necessary condition for an isotopy to remove a basic set Λ_i of f while not changing the other basic sets is that for any $m = 2^d r$, $2^{d+1} | N_m^i$ and if r is odd, $2^{d+1} \mid \sum_{s=0}^{d} K^i_{2^s r}$.
2. A necessary condition for an isotopy to "cancel" 2 basic sets Λ_i and Λ_j while leaving all others unchanged is the following (u_i is the index of Λ_i):
   a. if $u_i - u_j$ is even, then for any $m = 2^d r$, $2^{d+1} | (N_m^i + N_m^j)$
   b. if $u_i - u_j$ is odd, then for any $m = 2^d r$, $2^{d+1} | (N_m^i - N_m^j)$.
3. A necessary condition for an isotopy to substitute a basic set Λ_i for a basic set Λ_i' without altering the others is that:
   a. if $u_i - u_j$ is even, then for any $m = 2^d r$, $2^{d+1} | (N_m^i - N_m^{i'})$
   b. if $u_i - u_j$ is odd, then for any $m = 2^d r$, $2^{d+1} | (N_m^i + N_m^{i'})$.

BIBLIOGRAPHY

1. R. Bowen, Topological entropy and Axiom A, Proc. Sympos. Pure Math. 14, Amer. Math. Soc., Providence, R.I., 23-42.

2. J. Franks, A reduced zeta function for diffeomorphisms, to appear in Amer. J. of Math.

3. J. Franks, Some smooth maps with infinitely many hyperbolic periodic points, to appear in Trans. Amer. Math. Soc.

4. J. Guckenheimer, Axiom A and no-cycles imply $\zeta(f)$ rational, Bull. Amer. Math. Soc. 76(1970), 592-594.

5. A. Manning, Axiom A diffeomorphisms have rational zeta functions, Bull. London Math. Soc. 3(1971), 215-220.

6. C. Narasimhan, The periodic behavior of Morse-Smale diffeomorphisms on compact surfaces, Thesis, Northwestern University, 1977.

7. J. Neuberger, An iterative method for solving nonlinear partial differential equations, Advances in Math. 19(1976), 245-265.

8. M. Shub and D. Sullivan, Homology theory and dynamical systems, Topology 14(1975), 109-132.

9. S. Smale, Differentiable dynamical systems, Bull. Amer. Math. Soc. 73(1967), 747-817.

10. R. Williams, The zeta function of an attractor, Conference on Topology of Manifolds (Michigan State 1967), Prindle, Weber and Schmidt (1968).

EMORY UNIVERSITY

ENTROPY AND THE FUNDAMENTAL GROUP

by

Rufus Bowen*

Let $f: M \to M$ be a map of a compact manifold. We shall give two slight strengthenings of known results relating the entropy $h(f)$ of f to topological invariants.

Theorem 1. *Let* μ *be the growth rate of* f_* *on* $\pi_1(M)$. *Then* $h(f) \geq \log \mu$.

Theorem 2. *Let* $A: R^n \to R^n$ *be linear and* $f: R^n \to R^n$ *continuous with* $\|f-A\|_\infty < \infty$. *Then* $h(f) \geq h(A)$.

The first result is a simple reinterpretation of Manning's Theorem that $h(f) \geq \log |\lambda|$ for λ an eigenvalue of f_* on $H_1(M)$ [2]. Plykin's attractor [4] provides an example where $\mu > 1$ though all λ's are 1. The second result gives a covering space proof of a theorem of Misiurewicz and Przytycki [3] on maps of tori.

To make sense of Theorem 1 let us define the *growth rate* $GR(\alpha)$ of an endomorphism $\alpha: \Gamma \to \Gamma$ of a finitely generated group Γ. For $S = \{s_1, \ldots, s_n\}$ a set of generators of Γ one defines the length $L_S(\gamma)$ of $\gamma \in \Gamma$ to be the length of the shortest word in the letters $S \cup S^{-1}$ which represents γ. Define

$$GR(\alpha) = \sup_{\gamma \in \Gamma} \{\limsup_{m \to \infty} L_S(\alpha^m \gamma)^{1/m}\}.$$

One sees that it is enough to consider $\gamma \in S$ and that $GR(\alpha)$ is independent of the set S of generators used. For Γ a finitely generated abelian group one computes $GR(\alpha) = \max\{|\lambda_1|, \ldots, |\lambda_r|, 1\}$ where $\lambda_1, \ldots, \lambda_r$ are the eigenvalues of $\alpha \otimes 1$ on $\Gamma \otimes C$.

*Partially supported by National Science Foundation (MCS74-19388-A01).

Next we recall the definition of entropy for a continuous map $f: M \to M$ of a compact space. A set $E \subset M$ is said to (n,ε)-<u>span</u> for f if

$$\forall x \in M \ \exists y \in E \text{ so that } d(f^k x, f^k y) < \varepsilon \text{ for } k = 0,1,\dots,n-1.$$

By compactness, for any n and $\varepsilon > 0$ there is a finite set E which spans for f; let $r(n,\varepsilon)$ be the smallest cardinality of such a set. The entropy $h(f)$ is defined by

$$h(f) = \lim_{\varepsilon\to 0} \{\limsup_{n\to\infty} \frac{1}{n} \log r(n,\varepsilon)\}.$$

§1. Proof of Theorem 1

This is just a minor addition to Manning [2]. Pick some Riemannian metric on M and let $\ell(\gamma)$ denote the length of a rectifiable curve γ. Choose $x_0 \in M$ and ρ a path from x_0 to $f(x_0)$. Define f_* on $\pi_1(M) = \pi_1(M,x_0)$ by $f_*[\gamma] = [\rho\#\gamma\#\rho^{-1}]$ (Proposition 1(5) of Section 3 says that $GR(f_*)$ does not depend on the choice of x_0 or ρ.)

Lemma A [2]. <u>There is a</u> $\delta > 0$ <u>with the following property. If</u> $\gamma: [0,1] \to M$ <u>is a rectifiable curve and</u> $n \geq 0$, <u>then there is a curve</u> $\tilde{\gamma}: [0,1] \to M$ <u>so that</u>

(i) $\ell(\tilde{\gamma}) \leq (r(n,\delta)+2)([\frac{\ell(\gamma)}{\delta}]+1)$

(ii) $\tilde{\gamma}(0) = f^n\gamma(0)$, $\tilde{\gamma}(1) = f^n\gamma(1)$

<u>and</u>

(iii) $\tilde{\gamma} \sim f^n\gamma$ rel $\{0,1\}$.

Inductively one sees that

$$f_*^n[\gamma] = [\rho\#f\rho\#\dots\#f^{n-1}\rho\#f^n\gamma\#f^{n-1}\rho^{-1}\#\dots\#\rho^{-1}].$$

Lemma A says that this element of $\pi_1(M,x_0)$ contains a curve no longer than

$$([\frac{\ell(\gamma)}{\delta}] + 1)(r(n,\delta) + 2) + 2\ell(\rho) + 2([\frac{\ell(\rho)}{\delta}] + 1) \sum_{k=1}^{n-1} (r(k,\delta) + 2)$$

Now $h(f) \geq \limsup_{m\to\infty} \frac{1}{m} \log r(m,\delta)$ means that, for any $\beta > 0$, there is a constant c_β so that $r(k,\delta) \leq c_\beta e^{(h(f)+\beta)k}$ for all k. The length above is then less than

$$d_1 + d_2 n + d_3 \sum_{k=0}^{n} e^{(h(f)+\beta)k} \leq d_1 + d_2 n + d_4 e^{(h(f)+\beta)n}$$

where the d_j are constants depending on ρ, γ, and β.

It is standard that $L_S([\alpha]) \leq c\ell(\alpha)$ for some constant $c > 0$. Hence

$$L_S(f_*^n[\gamma]) \leq c(d_1 + d_2 n + d_4 e^{(h(f)+\beta)n}).$$

From this, $\limsup_{n\to\infty} \log(L_S(f_*^n[\gamma]))^{1/n} \leq h(f) + \beta$ and $\log GR(f_*) \leq h(f) + \beta$. Just let $\beta \to 0$ to finish.

§2. An Example

An example where the growth rates on π_1 and H_1 are different appears in Plykin [4]. Here M is the disk with 3 disks removed and f maps M into its interior with the following action on $\pi_1(M)$ (the free group on 3 generators).

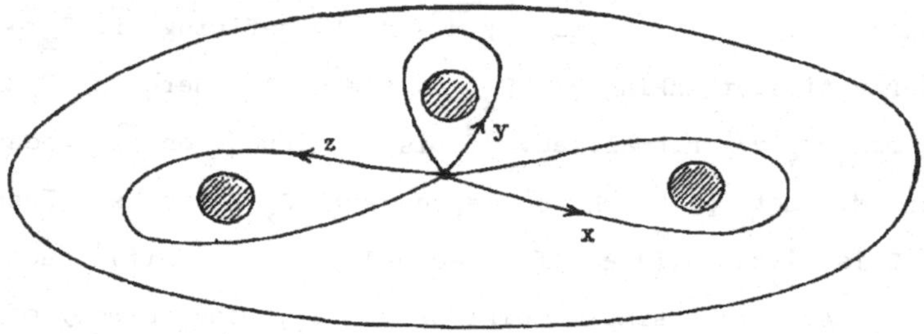

$$f_*(x) = zy^{-1}xyz^{-1}$$

$$f_*(y) = y^{-1}x^{-1}yxy$$

$$f_*(z) = zy^{-1}xyzy^{-1}x^{-1}yz^{-1}.$$

When one plugs these expressions into themselves repeatedly to calculate f_*^n, there is never any cancellation (the letters in the words so generated are alternately from the two sets $\{y,y^{-1}\}$ and $\{x,x^{-1},z,z^{-1}\}$). If one lets $x_1 = x$, $x_2 = y$, $x_3 = z$, then it follows inductively that the number of occurrences of x_j or x_j^{-1} in $f_*^n x_i$ equals $A^n_{i,j}$ where $A = \begin{pmatrix} 1 & 2 & 2 \\ 2 & 3 & 0 \\ 2 & 4 & 3 \end{pmatrix}$. The growth rate of the entries of A^n is the positive eigenvalue $\lambda > 1$ of A; hence $GR(f_*) = \lambda > 1$. On $H_1(M)$ one has f_* equal to the identity and growth rate 1.

Plykin extends f to $S^2 \supset M$ by adding a source in each of the four components of $S^2 \setminus M$. Let us restate the result in terms of this picture.

Proposition. <u>Let</u> $f\colon M \to M$ <u>be a diffeomorphism of a compact manifold and</u> X <u>be a finite set with</u> $f(X) = X$. <u>Then</u> $h(f) \geq \log \mu$ <u>where</u> μ <u>is the growth rate of</u> f_* <u>on</u> $\pi_1(M \setminus X)$.

<u>Proof</u>: Blow up each $x \in X$ to a sphere S_x. Let $\tilde{M}$ be the compact manifold (with boundary) thereby obtained; $\tilde{M}$ is the compactification of $M \setminus X$ where S_x is a boundary component where x was deleted. Now extend f on $M \setminus X$ to $\tilde{f}\colon \tilde{M} \to \tilde{M}$ by defining $\tilde{f}\colon S_x \to S_x$ via the derivative; thinking of S_x as the unit sphere in T_xM, let $\tilde{f}(v)$ be $Df_x(v)$ normalized. $\tilde{f}$ is continuous on $\tilde{M}$ because f was C^1 on M. Let $p\colon \tilde{M} \to M$ collapse each S_x onto x. Then $f \circ p = p \circ \tilde{f}$. This gives $h(\tilde{f}) \geq h(f)$; actually $h(\tilde{f}) = h(f)$ because each fibre $p^{-1}(y)$ is a simple point or an S_x, the entropy of f on one

of these fibres is 0 (the map on the sphere induced from any linear map has entropy 0). By Theorem 1, $h(\tilde{f}) \geq GR(f_* \text{ on } \pi_1(\tilde{M}))$. But $\pi_1(\tilde{M}) \cong \pi_1(M \setminus X)$ and this identification preserves f_*.

The proposition has content of course only for $\dim M = 2$, but the "blowing up" works for any dimension.

Thurston [5] has constructed canonical representatives for isotopy classes of diffeomorphisms of surfaces. The most interesting of these are the pseudo-Anosov diffeomorphisms f. Here there is an algebraic number λ which acts as a "stretching factor" everywhere except at a finite number of points. If one knew that $h(f) \leq \log \lambda$, then it would follow that $h(f) = \log \lambda$ and $GR(f_* \text{ on } \pi_1) = \lambda$. So f would attain the bound in Theorem 1.

§3. On Growth Rates

Proposition 1. Let $S = \{s_1, \ldots, s_n\}$ generate Γ and $\alpha: \Gamma \to \Gamma$ be an endomorphism. Then

(1) $GR(\alpha) = \lim_{m\to\infty} K_m^{1/m} = \inf_m K_m^{1/m}$ where $K_m = \max_{1\leq i\leq n} L_S(\alpha^m s_i)$

(2) $GR(\alpha^n) = GR(\alpha)^n$ for $n \geq 1$

(3) If H is a subgroup of Γ with $\alpha(H) \subset H$ and $[\Gamma:H] < \infty$, then $GR(\alpha|H) = GR(\alpha)$.

(4) If $\Gamma_1 = \Gamma$ and $\Gamma_{j+1} = [\Gamma,\Gamma_j]$, then

$$GR(\alpha) \geq GR(\alpha \text{ on } \Gamma_j/\Gamma_{j+1})^{1/j}.$$

(5) If $\beta(\gamma) = h^{-1}\alpha(\gamma)h$ for some $h \in \beta$, then $GR(\beta) = GR(\alpha)$.

Proposition 2. Suppose Γ is nilpotent (and finitely generated), i.e., $\Gamma_{t+1} = 0$ for some t. Then

$$GR(\alpha \text{ on } \Gamma) = \max_{0\leq k\leq t} GR(\alpha \text{ on } \Gamma_k/\Gamma_{k+1}).$$

Proof: We will prove that

$$GR(\alpha \text{ on } \Gamma) = \max \{GR(\alpha \text{ on } \Gamma/\Gamma_t), GR(\alpha \text{ on } \Gamma_t)^{1/t}\}.$$

The proposition will then follow by induction on t, applied to $\tilde{\Gamma} = \Gamma/\Gamma_t$. One has LHS $\geq$ RHS by (4) above and the fact that quotients can only have smaller growth rate. Let $K = GR(\alpha \text{ on } \Gamma/\Gamma_t)$ and $L = GR(\alpha \text{ on } \Gamma_t)$. Let $S = \{s_1,\ldots,s_n\}$ generate Γ, $\tilde{S} = \{s_1\Gamma_t,\ldots,s_n\Gamma_t\}$.

Fix m and write $\alpha^m(s_i) = w_i z_i$ where w_i is a word in $S \cup S^{-1}$ of length $\leq K_m = \max_{1\leq i\leq n} L_{\tilde{S}}(\alpha^m(s_i\Gamma_t))$ and $z_i \in \Gamma_t$. Let $M = \max_{1\leq i\leq n} L_T(z_i)$ where T is a set of generators for Γ_t and let $L_m = \max_{\gamma\in\Gamma} L_T(\alpha^m\gamma)$. By induction on j one sees that

$$\alpha^{jm}(\gamma_i) = w_{j,i}z_{j,i}$$

where $w_{j,i}$ is a word in $S \cup S^{-1}$ of length at most K_m^j and $z_{j,i} \in \Gamma_t$ with

$$L_T(z_{j,i}) \leq L_m^{j-1}M + L_m^{j-2}MK_m + \ldots + L_mMK_m^{j-2} + MK_m^{j-1}.$$

Here we use that Γ_t is in the center of Γ. By choosing S and T properly one can assume $L_S(z) \leq cL_T(z)^{1/t}$ for all $z \in \Gamma_t$ and some constant c (see Wolf [6], statement 3.8 on p. 429). Hence

$$\begin{aligned} L_S(\alpha^{jm}s_i) &\leq K_m^j + c(L_m^{j-1}M + L_m^{j-2}MK_m + \ldots + MK_m^{j-1})^{1/t} \\ &\leq K_m^j + c(jM)^{1/t}\max\{L_m^{j/t}, K_m^{j/t}\}. \end{aligned}$$

Since $GR(\alpha) = \lim_{j\to\infty} \max_{1\leq i\leq m} L_s(\alpha^{jm}s_i)^{1/jm}$, one has

$$GR(\alpha) \leq \max\{K_m^{1/m}, L_m^{1/tm}\}.$$

We are finished by (1) of Proposition 1.

In general calculating the growth rate does not seem easy.

Thurston's work (see remarks at end of Section 2) probably contains a proof that automorphisms of surface groups have algebraic growth rates. One wonders how generally this is true.

§4. Proof of Theorem 2

For a map $f: R^n \to R^n$ one must modify the definition of entropy $h(f)$ because R^n is not compact [1]. For $K \subset R^n$ compact a subset $E \subset K$ will (n,ε)-span K (for f) if

$$\forall\, x \in K \;\; \exists\, y \in E \text{ so that } \|f^k x - f^k y\| < \varepsilon \quad \text{for } k = 0,1,\dots,n-1.$$

Let $r(n,\varepsilon,K)$ be the smallest cardinality of such a set. Define

$$h(f,K) \;=\; \limsup_{n\to\infty} \frac{1}{n} \log r(n,\varepsilon,K)$$

and

$$h(f) \;=\; \sup\,\{h(f,K):\; K \subset R^n \text{ compact}\}.$$

The number $h(f)$ does not depend on which norm is used on R^n.

For $A: R^n \to R^n$ linear one has $h(A) = \sum_{|\lambda_i|>1} \log |\lambda_i|$ where $\lambda_1,\dots,\lambda_n$ are the eigenvalues of A [1]. Let $\alpha_1,\dots,\alpha_m$ be the distinct norms of eigenvalues $|\lambda_i|$ greater than 1, and let V_j be the A-invariant subspace of R^n corresponding to all λ_i's with $|\lambda_i| = \alpha_j$. Let $V = \oplus_{j=1}^m V_j$ and W be the invariant subspace corresponding to all eigenvalues with $|\lambda_i| \le 1$, so that $R^n = V \oplus W$. Let $\pi: R^n \to V$ denote projection along W. Fix $\varepsilon > 0$ and norm each V_j so that

$$\|Av\| \;\ge\; (\alpha_j-\varepsilon)\|v\| \qquad \text{for all } v \in V_j.$$

Put any norm on W and the sup norm on $V = \oplus V_j$ and $R^n = V \oplus W$.

Let $a = \sup_{x\in R^n} \|f(x) - A(x)\| < \infty$, $b > 0$ to be chosen later and

$$B_k \;=\; \{\textstyle\sum v_j \in V:\; \|v_j\| < b(\alpha_j-2\varepsilon)^k \text{ for all } j\}.$$

The homology group $H_s(V \setminus B_k) \cong Z$ has ∂B_k for a generator where $s = \dim V - 1$. Now suppose ε is small enough that $\min_j (\alpha_j - 2\varepsilon) > 1$ and b so large that $b\varepsilon > a$. Then one has $A(V \setminus B_k) \subset V \setminus B_{k+1}$, $f((V \setminus B_k) \oplus W) \subset (V \setminus B_{k+1}) \oplus W$, and a commutative diagram

$$\begin{array}{ccc} H_s((V \setminus B_k) \oplus W) & \xrightarrow{f_*} & H_s((V \setminus B_{k+1}) \oplus W) \\ \downarrow \pi_* & & \downarrow \pi_* \\ H_s(V \setminus B_k) & \xrightarrow{A_*} & H_s(V \setminus B_{k+1}) \end{array}$$

Each of these groups is infinite cyclic and each homomorphism an isomorphism. It follows that $\pi f^k(\partial B_0 \oplus 0)$ and $A^k(\partial B_0)$ represent the same class in $H_s(V \setminus B_k)$, namely a generator. Now $\pi f^k(B_0 \oplus 0) \supset B_k$; for if $p \in B_k \setminus \pi f^k(B_0 \oplus 0)$, then the sequence of maps

$$\partial B_0 \oplus 0 \subset B_0 \oplus 0 \xrightarrow{\pi f^k} V - \{p\} \to V \setminus B_k$$

on homology would send $\partial B_0 \oplus 0$ to a generator of $H_s(V \setminus B_k) \cong Z$ even though $H_s(B_0 \oplus 0) = 0$.

Let $E_k \subset B_0 \oplus 0$ be a set which $(k+1, \varepsilon)$-spans $B_0 \oplus 0$ with respect to f. Then

$$\pi f^k(B_0 \oplus 0) \subset \bigcup_{y \in E_k} \{\varepsilon\text{-ball about } \pi f^k(y)\}$$

and so $\mathrm{vol}(\pi f^k(B_0 \oplus 0)) \leq \text{constant} \cdot \mathrm{card}(E_k)$. Because $\pi f^k(B_0 \oplus 0) \supset B_k$ we have

$$\mathrm{card}(E_k) \geq \text{constant} \prod_j (\alpha_j - 2\varepsilon)^{k \cdot \dim(V_j)}.$$

Letting $k \to \infty$ this implies

$$h(f) \geq h(f, B_0 \oplus 0) \geq \sum_{|\lambda_i| > 1} \log(|\lambda_i| - 2\varepsilon).$$

Now let $\varepsilon \to 0$.

If $f\colon T^n \to T^n$, then f is homotopic to a map of T^n induced by a linear map $A\colon R^n \to R^n$ with $A(Z^n) \subset Z^n$. The map f lifts to $\tilde{f}\colon R^n \to R^n$ with $\|f - A\|_\infty < \infty$ and A can be identified with f_* on $H_1(T^n,R) \simeq R^n$. As $h(\tilde{f}) = h(f)$, Theorem 2 says that $h(f) \geq \sum_{|\lambda_i|>1} \log |\lambda_i|$ where $\lambda_1,\ldots,\lambda_n$ are eigenvalues of f_* on $H_1(M,R)$, a result of Misiurewicz and Pryzytycki [3]. For a nilmanifold $M = N/\Gamma$ one can lift $f\colon M \to M$ to $\tilde{f}\colon N \to N$. One can project $N \to N/[N,N] \simeq R^q \to V$ and the proof of Theorem 2 will give $h(f) \geq \sum_{|\lambda_i|>1} \log |\lambda_i|$ where $\lambda_1,\ldots,\lambda_n$ again are the eigenvalues of f_* on $H_1(M,R) \simeq N/[N,N]$. For nilmanifolds there is a better inequality which Manning has announced: if A is the algebraic endomorphism of N homotopic to f, then $h(f) \geq h(A)$.

Remark. Since writing this note we have found that Katock ("The Entropy Conjecture" in "Smooth Dynamical Systems" (D. Anosov, edit.; in Russian)), Manning, and Shub all thought of Theorem 1 too.

REFERENCES

1. R. Bowen, Entropy for group endomorphisms and homogeneous spaces, Trans. Amer. Math. Soc. 153(1971), 401-414 and 181(1973), 509-510.
2. A. Manning, Topological entropy and the first homology group, Springer Lecture Notes in Math. 468, 185-190.
3. M. Misiurewicz and F. Przytycki, Entropy conjecture for tori, preprint.
4. R. Plykin, Sources and sinks for A-diffeomorphisms, Math. USSR Sbornik 23(1974), 233-253.
5. W. Thurston, On the geometry and dynamics of diffeomorphisms of surfaces, preprint.
6. J. Wolf, Growth of finitely generated solvable groups and curvature of Riemannian manifolds, Jour. Diff. Geom. 2(1968), 421-446.

UNIVERSITY OF CALIFORNIA
AT BERKELEY

ISOLATED INVARIANT SETS OF PARAMETERIZED SYSTEMS OF DIFFERENTIAL EQUATIONS

by

C. Conley and Joel Smoller

§1. The Problem

The question of existence of traveling wave solutions to non-linear diffusion reaction equations frequently reduces to the study of special solutions of a parameterized family of ordinary differential equations; namely, solutions which run from one critical point to another (e.g., see [1]). A typical such problem is represented by the following simple example, in which the parameter, λ, ranges over the interval from -1 to 1.

$$\dot{x} = xy - \lambda/10$$
$$\dot{y} = 1 - y^2.$$

Phase portraits for the values $\lambda = -1$, $\lambda = 0$ and $\lambda = +1$ are indicated in Figure 1 a, b and c respectively. The large rectangle,

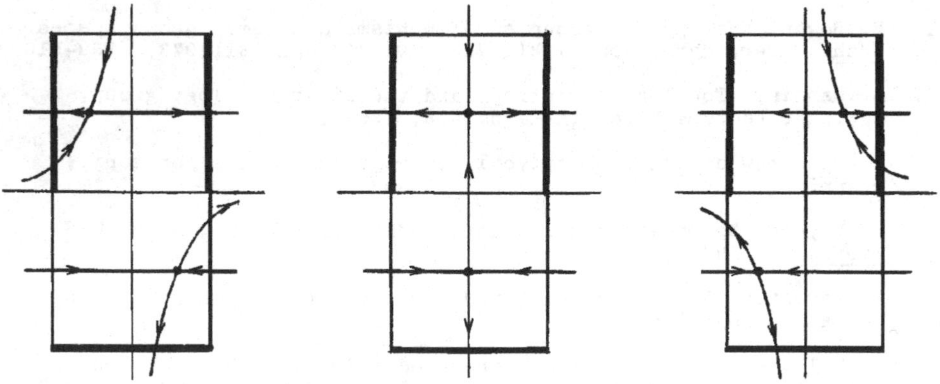

(a) (b) (c)

Figure 1

which is divided into two squares, will be discussed later; for now it is only noted that there are two rest points in each portrait; these are given by $y = \pm 1$ and $x = \lambda/10y$. They lie in the second and fourth quadrants if $\lambda < 0$ and in the first and third if $\lambda > 0$. In all three portraits the lines $y = \pm 1$ are invariant and correspond to the stable $(y = -1)$ or unstable $(y = +1)$ manifold of the corresponding rest point.

The first and third portraits differ in the way the other stable and unstable manifolds leave the large rectangle, and this difference clearly has something to do with the solution connecting the two rest points in the middle portrait.

This connecting solution is obviously unstable in the sense that if one slightly perturbs the system $\dot{x} = xy$; $\dot{y} = 1 - y^2$ then it is (generically) expected that the connection will disappear. On the other hand, if one perturbs the family of equations (maintaining the continuous dependence on λ of course) the behavior at $\lambda = \pm 1$ still should force the existence of a connecting solution for some parameter value in the interval from -1 to 1. The aim here is to introduce an "index" for families of equations which allows one to see that this is the case.

§2. The Index of an Isolated Invariant Set

The index we have in mind is something like the index of an isolated invariant set (cf. [2], [3], [4] and [5]) which will now be described. An invariant set (i.e., a union of orbits) is called isolated if it admits a compact neighborhood such that any orbit in the neighborhood is also, in the set. Such a neighborhood is called an isolating neighborhood. For example, in Figure 1b there are four distinct non-empty isolated invariant sets: the lower rest point, the upper rest point, the union of these two sets and, finally, the two points together with the connecting solution. Isolating neighborhoods are provided (respectively) by the lower square, the upper square, the union of two slightly smaller squares concentric with the two mentioned (and not shown) and, finally, the large rectangle.

In the first two and the last of these examples the isolating neighborhoods have a special property; namely, any boundary point of the neighborhood leaves the neighborhood immediately either as time increases or as time decreases. Neighborhoods like these are called isolating blocks and can be used to compute the index of the isolated invariant set inside.

The index is the homotopy type of the pointed space obtained from the block on collapsing to one point all points which leave the block immediately in forward time (i.e., the "exit points;" this set of exit points must lie in the boundary of the block of course). For example, in Figure 2a, the block for the lower rest point is the lower square and the two segments of exit points are heavily drawn. On collapsing

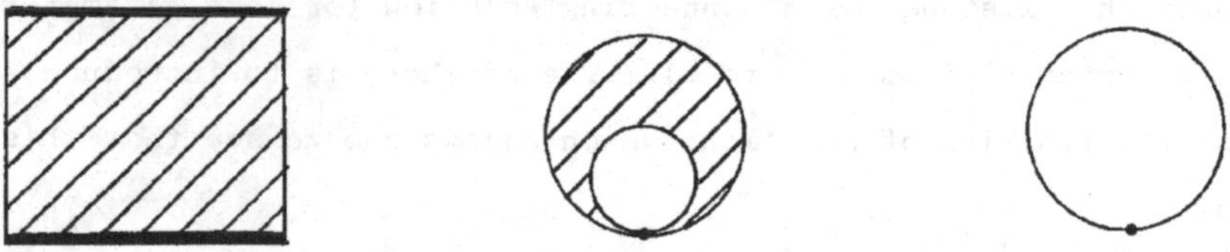

(a) (b) (c)

Figure 2

these two segments to a point one has a space somewhat like that in Figure 2b. This space is homotopic to ("can be deformed to") a pointed circle as shown in Figure 2c. Thus the index of the lower rest point is the homotopy type of a pointed circle -- which is denoted here by Σ^1. Of course this definition requires that any other block would give the same result; there is a theorem to that effect. In a similar way one can see that the index of the upper rest point is also Σ^1.

The index of the disjoint union of two isolated invariant sets is obtained by identifying the distinguished points of the two indices to one point; the resulting space is called the sum of the given two. To prove the sum formula, one chooses disjoint blocks for the given invariant sets. Their union is then a block for the union of the sets. The exit set of the union is the union of the exit sets and the result becomes clear on collapsing it to a point. For example, the index of the union of the two rest points is the homotopy type of the (pointed) figure eight shown in Figure 3. This is denoted $\Sigma^1 \vee \Sigma^1$, $\vee$ being the symbol for "addition." This is also the index of the set consisting of the two rest points and the connecting orbit, as can be seen by collapsing out the three heavily shaded segments which comprise the exit set of the rectangular block in Figure 1b (see Figure 4; this also gives an alternate computation of the index of the union of the two points in Figure 1a or 1c).

Figure 3

Now the empty set is also an isolated invariant set: an isolating neighborhood as well as an isolating block is provided by the empty set itself. Furthermore the exit set of the block is empty and on collapsing this set to a point one obtains the pointed one point space. The homotopy type of this space, hence the index of the empty set, is denoted $\overline{0}$. Observe that $\overline{0}$ is the additive identity, which is compatible with the sum formula.

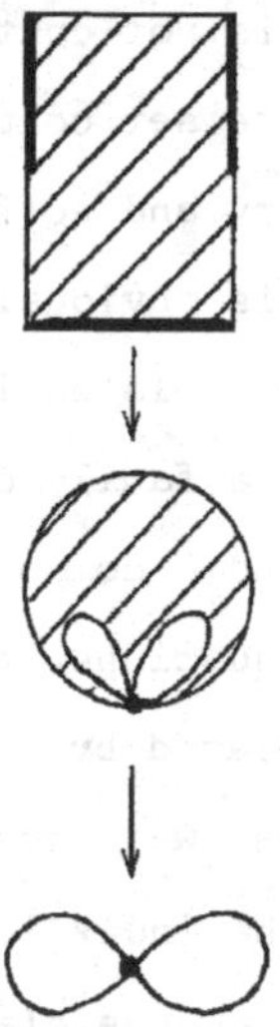

Figure 4

To see how this index works in existence proofs, suppose given any equation (in the plane) which admits a block like the rectangle in Figure 1 (in particular each boundary point must leave either in forward or in backward time and those leaving in forward time make up the three heavily shaded arcs). Then the isolated invariant in the block has index $\Sigma^1 \vee \Sigma^1$. Since $\Sigma^1 \vee \Sigma^1$ is not the same as $\overline{0}$, the invariant set inside the rectangle cannot be empty.

This conclusion could also be derived from a winding number argument which would show that the sum of the Euler indices of the critical points in the rectangle is -2. In general, however, the above index does not imply the existence of critical points -- for example, a hyperbolic periodic orbit in R^3 with two dimensional stable and two dimensional unstable manifold is an isolated invariant set with non-zero index.

§3. Continuation

If N is a compact set such that the orbit through any boundary point of N is not contained in N, then N is an isolating neighborhood for the set comprised of the points on orbits contained in N. This (necessary and sufficient) condition that N be an isolating neighborhood is obviously "stable" to small perturbations and allows one to relate isolated invariant sets of different equations. Specifically, given a family of equations parameterized by an arc (e.g., by $\lambda \in [-1,1]$) and a compact set, N, which is an isolating neighborhood for <u>all</u> the equations, the corresponding isolated invariant sets are said to be related by continuation. Making the relation transitive (in the minimal way) one has an equivalence relation on isolated invariant sets. The index is the same for all isolated invariant sets in the same equivalence class.

For example, in Figure 1 each of the two squares and also the larger rectangle are isolating neighborhoods for all values of λ.

Therefore all of the lower rest points, say, are related by continuation; similarly all the upper ones are.

The rectangle determines, for each $\lambda \neq 0$ the isolated invariant set consisting of the two rest points, and, for $\lambda = 0$, that consisting of the two rest points together with the connecting orbit. Thus all of the sets in this family are in the same equivalence class so have the same index (as has already been seen).

On replacing the rectangle by the union of two slightly shrunken squares one determines another family of related sets which agrees with that above except for $\lambda = 0$ when the set determined consists only of the two rest points (without the connecting orbit). In particular, this shows that different invariant sets in the same equation can be related by continuation.

It is these last two families of isolated invariant sets that we wish to distinguish in terms of some sort of index. The present one doesn't do the job since there is no difference in the "invariant set to index" correspondence; of course the present index does not take into account the observed difference between $\lambda = -1$ and $\lambda = +1$.

§4. Attractor-Repeller Pairs

Suppose S is an invariant set. An attractor, A, in S is an invariant set in S which admits a strictly positively invariant neighborhood, B, in which it is the maximal invariant set. That B is strictly positively invariant means that (in terms of the topology on B relative to S), each boundary point of B goes immediately to the interior of B as time increases (in particular, relative to S, A is an isolated invariant set and B is a block for A with empty exit set).

If S is compact, then to each attractor A. there corresponds a unique repeller (i.e., attractor for reverse time) A^* defined as the complement in S of the set of points whose orbit tends to A in forward time. A generic attractor-repeller pair is labeled (A,A^*). The set $S\setminus A\cup A^*$ will be called $C(A,A^*)$; all points in this set tend to A^* as time goes to $-\infty$ and of course, to A as time goes to $+\infty$.

An example is provided by the isolated invariant set in the rectangle in Figure 1b; S is the segment, A is the upper rest point, A^* is the lower rest point and $C(A,A^*)$ is the connecting orbit; in particular, $C(A,A^*) \neq \emptyset$. Another example is provided by the related set for any of the other equations; A and A^* can be chosen to be the same while S is now $A \cup A^*$ and $C(A,A^*) = \emptyset$. (The positively invariant set B corresponding to A is A itself, which is both open and closed in this S).

In this last example it is important to observe that the roles of A and A^* can be reversed: in general any closed open subset of S is both an attractor and a repeller relative to S and its complement in S is the dual repeller or attractor as the case may be. However, the roles of A and A^* cannot be reversed in the first example because A^* is definitely not an attractor in S.

Now it is generally true that attractor-repeller pairs in S continue (under perturbation) to attractor-repeller pairs in the continuation of S. However, as happens above when λ changes from 0, the connection might continue to the empty set.

In the present problem, though, the connection is only expected to remain under perturbation in the sense that, when the family is perturbed to a new family, a connecting orbit exists for some value of λ. This suggests that λ be treated as an additional variable satisfying the equation $\dot{\lambda} = 0$ so that the problem can be considered as involving just one system of equations. The state space of the equation is considered to be the product of the closed interval $[-1,1]$ and the x,y-plane. One now finds that an individual rest point is no longer an isolated invariant set. However, the full set of upper rest points comprise an isolated invariant set, as does the set of lower rest points, the union of these two sets, and the union of this set with the one connecting orbit. (These statements would not be true if the closed interval were replaced with an open one because the required compactness would be lost.) Again the first and second of these sets form an attractor-repeller pair in the last and the connection between them is non-empty. The problem is to show that, under perturbation, this connection remains non-empty.

§5. A Modified Equation

In order to bring in the different behaviors at $\lambda = \pm 1$ the equations will now be modified somewhat; namely, the equation $\dot{\lambda} = 0$ will be changed to $\dot{\lambda} = \varepsilon\lambda(1-\lambda^2)\varphi(x,y)$ where ε is a small positive parameter. The function $\varphi(x,y)$ is defined on a neighborhood (in $[-1,1] \times R^2$) of the product of the rectangle with the interval $[-1,1]$. It is positive in a small tube about the set of lower rest points and negative in a tube about the set of upper rest points and is zero otherwise. The effect of this change in the λ-equation is indicated in Figure 5 (which holds for any $\varepsilon > 0$): now there are only six rest points left, those former ones for which $\lambda = -1$, 0 or 1. The important ones are those with $\lambda = 0$.

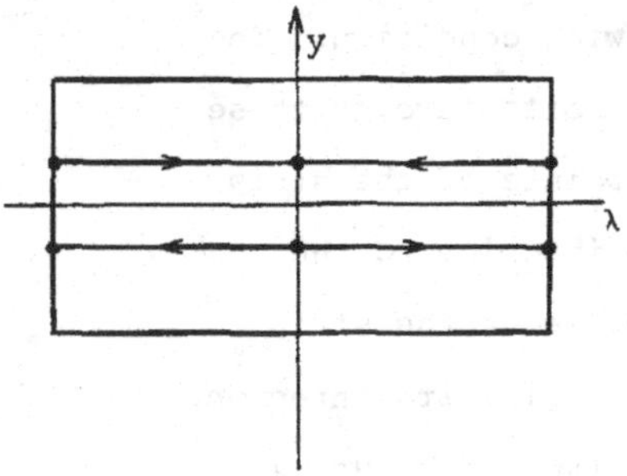

Figure 5

The stable manifold of the upper one of these is two dimensional, as is the unstable manifold of the lower one; furthermore the intersection of these manifolds is obviously stable. In fact it can be "measured" by the index introduced above as will now be described.

The product of the rectangle (in Figure 1) with the interval $[-1,1]$ is an isolating neighborhood (even a block) for the three dimensional equation (note that the boundary of this neighborhood does not include the product of the interior of the rectangle with the end points of the interval -- the total space is $[-1,1] \times R^2$). The isolated invariant set in this block consists of the six rest points and (possibly) solutions connecting them as indicated in Figure 5. (The existence of the "horizontal" connecting solutions is not important for the present purpose and will not be verified. The subset of this isolated invariant set consisting of the rest points with $\lambda = 0$ together with the orbit connecting them is actually the set of interest. We would like to judiciously remove pieces of the block just described in order to get one for this subset so that its index can be computed.

There will be one piece for each rest point in the ends. First consider the upper left hand point; the appearance of its stable set in the rectangle $\lambda = -1$ (shown in Figure 1a) is shown again in Figure 6 together with a strip-like neighborhood satisfying the following condition. The positive orbits through those boundary points of the strip which are interior to the rectangle, leave the strip without entering its interior. It is constructed by using orbit segments for the boundaries until close enough to the unstable manifold of the rest point to ensure success in satisfying the condition (cf. Figure 6). Of course a similar strip can be constructed about the unstable manifold of the upper rest point

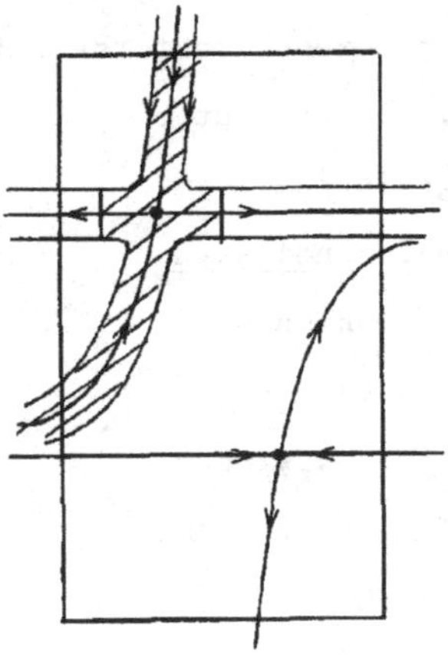

Figure 6

in the rectangle $\lambda = 1$. Furthermore, such strips can be constructed about the unstable manifolds of the lower rest points in either end of the block except it is now the negative half-orbit that is not allowed to enter the interior of the strip.

Having constructed such strips, it is clear that they can be modified slightly so that each boundary point of the strip interior to the rectangle leaves the strip immediately as time increases (or decreases in the case of the lower rest points).

Now observe that the stable manifold to the upper rest points in the end rectangles is likewise the stable manifold of these points relative to the full three dimensional block (cf. Figure 5); also the unstable manifold to the lower points in these rectangles is the stable manifold relative to the full block. Therefore, the strips can be replaced by (half-) tubes which satisfy the analogous boundary behavior with respect to the full three-dimensional block. On removing these tubes (or rather their interiors relative to the large state space) one is left with the desired block for the two middle rest points and their connecting orbit. (The procedure described above, i.e., removing pieces of the block to obtain a block about the smaller invariant set, is an example of a quite general one which is used to prove the abstract theorem typified by this example.)

It is now possible to compute the index of the invariant set containing the two rest points and the connecting orbit, and it is at this point that the difference in the flows at $\lambda = -1$ and $\lambda = +1$ plays a crucial role. A picture of the block just described is drawn in Figure 7. It is not quite accurate in that the indentations due to the removal of the tubes are not shown. Also it is distorted somewhat so that the top, bottom and both ends are shown; this allows the exit set to be seen more clearly.

The exit set can be described as the union of five pieces. First are the three pieces which come from the product of the interval

[-1,1] with each of the three exiting arcs in the rectangle(s) of Figure 1. The remaining two pieces lie one in each end of the block and have somewhat the aspect of the strip about the unstable manifold of the lower rest point. Of course these shaded strips should be the closures of the sets of those boundary points of the half-tubes which lie interior to the original block. It is easy to check that there are no more exit points (the boundary points of the half-tubes about the stable manifold of the upper rest points are entrance points, not exit points).

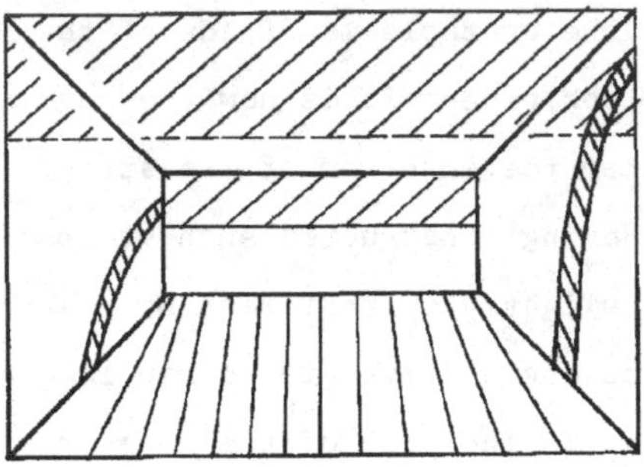

Figure 7

Having done this a representative of the index is obtained by collapsing the exit set to one point. Observe that (due to the difference in the equations at $\lambda = -1$ and $\lambda = +1$) this exit set is contractible to a point in the block so the resulting space is just the pointed three ball. Since the ball is also contractible, this has the homotopy type of the one point space; that is, the index is $\overline{0}$ (the same as that of the empty set).

It is important to note that the above construction can be carried out for any positive ε and the corresponding isolated invariant sets are related by continuation.

§6. An Existence Proof

Before defining the index for this family, one can see that the above construction leads to a proof of existence of the connecting orbit (after perturbation). This requires knowledge of the indices of the two rest points in the set whose index has just been computed.

Both of these rest points are hyperbolic points and, by a general result, each is therefore an isolated invariant set whose index is a pointed sphere, the dimension of which is that of the unstable manifold. Thus the index of the lower point is Σ^2 (the pointed two-sphere), and that of the upper point is Σ^1.

Suppose now that the original family of equations are perturbed somewhat. The argument that a connection persists goes as follows.

First, the construction of modified equations can be carried out as before and the blocks constructed in the unperturbed case are at least isolating neighborhoods for the new case. Therefore, by the invariance of the index under continuation, for each $\varepsilon > 0$ the set with index $\bar{0}$ continues to one with index $\bar{0}$ and the rest points in it continue to isolated invariant sets with indices Σ^1 and Σ^2 (respectively).

Now, remembering the sum formula and noting that $\Sigma^1 \vee \Sigma^2 \neq \bar{0}$, it follows that <u>the invariant set contains more than just the two rest points</u>. Since the continuations of the rest points form an attractor-repeller pair in the continued set, the additional orbits must connect them.

Observing that the maximal invariant set in the block depends upper-semi-continuously on ε, it follows that when $\varepsilon = 0$, there is still some solution which runs from the lower critical point family to the upper one. This completes the description of the proof.

§7. "Counter"-Examples

In the previous argument a crucial step was to modify the given equations and thereby create an isolated invariant set whose index (along with the sum formula) allowed one to conclude that equations arbitrarily near the given one had solutions which connected the arcs of rest points. The different behavior of the given equations at $\lambda = -1$ and $\lambda = +1$ determined that the index of the created invariant set was $\overline{0}$. If the behavior were the same at $\lambda = +1$ as it is at $\lambda = -1$ the block in question would appear as in Figure 8. Up to homotopy, it is a three ball and its exit set consists of a circle and a point. The index is then $\Sigma^1 \vee \Sigma^2$, which is the sum of those of the critical points. Therefore one cannot conclude the existence of other solutions.

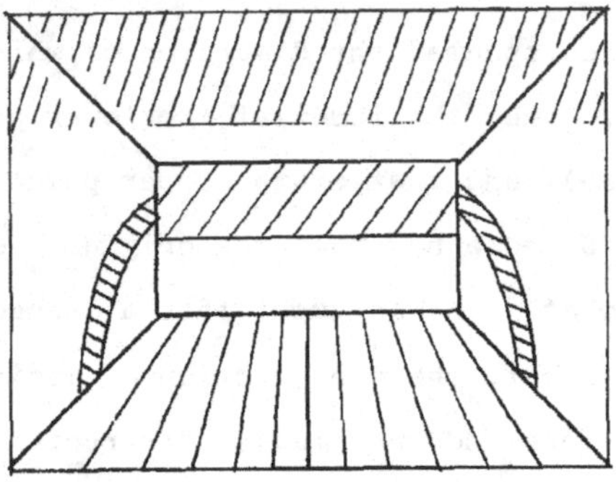

Figure 8

Also, in the example, a choice was made at the beginning; namely, to prove there was a connection from the lower point to the upper one (it has not been pointed out, but the indices Σ^1 and Σ^2 for the critical points and $\overline{0}$ for the large set automatically imply that the connection goes from the lower (Σ^2) point to the upper one). Had we tried to go the other way, the ε in the modified equation would have been chosen negative and the roles of the stable and unstable manifolds of the upper and lower points would have been reversed. In this case the index of the created invariant set would again have been $\Sigma^1 \vee \Sigma^2$. (This will be easier to see after reading the next paragraph).

§8. A "Relativized" Index

In the general set up there are given a family of equations parameterized by a compact space Λ (e.g., $[-1,1]$) and a corresponding family of isolated invariant sets, $\{S_\lambda \mid \lambda \in \Lambda\}$, all of which are related by continuation.

There are also given a compact subset, Λ_0, of Λ (e.g., $\{-1,+1\}$) and a corresponding family of attractor-repeller pairs, $\{(A_\lambda, A_\lambda^*) \mid \lambda \in \Lambda_0\}$, in S_λ. In the example, the pair is (lower rest point, upper rest point), <u>not</u> the other way around, although that too is an attractor repeller pair.

It is also necessary that Λ_0 be an attractor for some equation on Λ in order to be sure that the relativized index can be defined. (Note that $\{-1,1\}$ is an attractor for the equation $\dot{\lambda} = -\lambda(1-\lambda^2)$ on $[-1,1]$).

Defining $S = \cup \{S_\lambda \mid \lambda \in \Lambda\}$ and $A_0 = \{A_\lambda \mid \lambda \in \Lambda_0\}$, the above data is represented by (S,A_0). The general result states that the data (S,A_0) determines an index ($\bar{0}$ in the example) which is invariant under continuation (perturbation).

The index is defined as follows: one chooses a block B for the isolated invariant set S of the "extended" equation obtained on adding the equation $\dot{\lambda} = 0$ to the given system. Now for each $\lambda \in \Lambda_0$ there is a set of points in B which tend to A_λ in backward time (these orbits cannot be in S; otherwise A_λ would not be an attractor in S_λ). Let $\bar{B}^-$ be the union of all these "unstable manifolds" together with the exit set (usually denoted B^-) of B.

The index $h(S,A_0)$ determined by the data (S,A_0) is then the homotopy type of $B/\bar{B}^-$. Of course there is a little work to do in order to see that this index is independent of the choice of B. The construction in §5 indicates this is done by realizing the relative index as an ordinary one.

The force of the statement that the index is invariant under continuation is this: for small perturbations, the family S over Λ continues to a similar family, S', and the family A_0 over Λ_0 continues also to a new family of attractors A_0'. Then $h(S',A_0') = h(S,A)$. In fact an equivalence relation on such pairs (S,A_0) can be defined just as it is for isolating invariant sets.

§9. The "Trivial" Case

In the present context, the "trivial" case is that where the attractor-repeller pairs (A_λ,A_λ^*) continue (as such) over all of Λ. It doesn't seem easy to get a general theorem about what the relativized index would be in this case; however, if Λ is contractible then a general theorem is available.

In order to state the result it is necessary to introduce the "smash" product of (homotopy types of) pointed spaces. If X and Y are such spaces, with distinguished points x_0 and y_0 respectively, then $X \wedge Y$ is the pointed space $X \times Y / (X \times \{y_0\} \cup \{x_0\} \times Y)$. This product is well-defined on homotopy classes.

The product above arises in the following way: given two equations, say one in x-space and one in y-space, one can consider these two equations as one equation on (x,y)-space. Then if S_x is an isolated invariant set of the first and S_y of the second, $S_x \times S_y$ is an isolated invariant set of the "product" equation.

If B_x and B_y are isolating blocks for S_x and S_y (respectively) then $B_x \times B_y$ is an isolating block for $S_x \times S_y$. The exit set of this block is $B_x \times B_y^- \cup B_x^- \times B_y$ and it easily follows that $h(S_x \times S_y) = h(S_x) \wedge h(S_y)$.

Coming back to the computation of the relative index in the trivial case where (A_λ, A_λ^*) continues over Λ, we next recall that Λ_0 is (by assumption) an attractor for some equation on Λ. The index of the corresponding repeller, Λ_0^*, is $h(\Lambda_0^*) = [\Lambda/\Lambda_0]$. This repeller is, of course, an attractor for the reverse flow and the index of this attractor is denoted $h^*(\Lambda_0^*)$.

Assuming Λ is contractible, the relative index, $h(S, A_0)$, is then given by $\{h(A_\lambda) \wedge h(\Lambda_0^*)\} \vee \{h(A_\lambda^*) \wedge h^*(\Lambda_0^*)$. A feeling for why this should be true comes from inspection of Figure 5: in this figure, it appears, for the modified flow, that the upper rest point -- A_λ^* in the example -- corresponds to the product of the <u>attractor</u> Λ_0^* with A_λ^* while the lower rest point -- A_λ in the example -- corresponds to the product of the <u>repeller</u> Λ_0^* with A_λ.

In the example, $h(A_\lambda) = h(A_\lambda^*) = \Sigma^1$; also $h(\Lambda_0^*) = \Sigma^1$ (this is the index of a repelling point in the real line) while $h^*(\Lambda_0^*)$ in Σ^0

(the pointed, two point space). Therefore, if (A_λ, A_λ^*) were to continue over the whole interval as an attractor-repeller pair, then $h(S, A_0)$ would be equal to $\{\Sigma^1 \wedge \Sigma^1\} \vee \{\Sigma^1 \wedge \Sigma^0\}$. Since $\Sigma^n \wedge \Sigma^m$ is, in general Σ^{n+m}, this is the same as $\Sigma^2 \vee \Sigma^1$ (as has already been pointed out in Section 6).

Of course, in the example $h(S, A_0)$ has been already computed to be $\overline{0}$ so it is again seen that (A_λ, A_λ^*) does not continue over Λ.

§10. Concluding Remarks

The main aim of these remarks has been to see that given the families $\{S_\lambda \mid \lambda \in \Lambda\}$ and $\{(A_\lambda, A_\lambda^*) \mid \lambda \in \Lambda_0\}$, an "index" $h(S, A_0)$ can be defined which gives some information about the possiblity of continuing the second family over Λ. One application is envisaged wherein the impossibility of the continuation implies the existence of a special solution of the given differential equation.

In order to make the conclusion that the second family does not continue, it is also necessary to have some alternate way of computing $h(S, A_0)$ under the assumption that it does. We have given one only in the case where Λ is contractible.

The example treated is a simple one and the existence statement given there can be done much more easily. However, the following remarks give some indication of the value of this index approach.

Let $\dot{z} = g(z)$ be an equation on R^n (i.e., $z \in R^n$) with an isolated invariant set S' such that $h(S') \neq \overline{0}$. Now consider the (parameterized) system on R^{2+n} composed of the equations

$$\dot{x} = xy - \lambda/10$$

$$\dot{y} = 1 - y^2$$

$$\dot{z} = g(z).$$

Replace the previous families $\{S_\lambda \mid \lambda \in \Lambda\}$ and

$\{(A_\lambda, A_\lambda^*) \mid \lambda \in \Lambda_0\}$ by $\{S' \times S_\lambda \mid \lambda \in \Lambda\}$ and $\{(A_\lambda \times S', A_\lambda^* \times S') \mid \lambda \in \Lambda_0\}$ and, to save notation, denote the latter families by S_λ and (A_λ, A_λ^*) again.

Then (the new) $h(S, A_0)$ is still $\overline{0}$ (it is actually $\overline{0} \wedge h(S')$, but for any pointed space, X, $\overline{0} \wedge X = X \wedge \overline{0} = \overline{0}$). Also, if the attractor-repeller pair were to continue over Λ, this relative index would have to be $h(S')$ times the old one.

Now it is not too much to assume this latter index is not zero (the product of non-zero indices could be zero) so that it can again be concluded that the attractor-repeller pair does not continue over Λ. Of course this is obvious for the product equation; the point is that it is still true for nearby equations by the continuation theorem for the index.

In fact, it is true for any family of equations which can be connected to the given one along an arc in such a way that the family, S_λ over Λ continues on the arc and the family (A_λ, A_λ^*) over Λ_0 continues along the arc. We might say that the example is a model for a certain class of connection problems; namely those that can be reached in the above way (product plus continuation) from the example.

Of course one would generally think of going backwards: given a connection problem, can it be reduced to the example treated here. Some such examples will be discussed in later papers wherein the proofs omitted here will also appear. Also, some other model problems will be mentioned.

BIBLIOGRAPHY

1. Fife, P., Asymptotic states for the equations of reaction and diffusion, (to appear in Bull. Amer. Math. Soc.).

2. Conley, C.C., Isolated Invariant Sets and the Morse Index, (Lectures from a C.B.M.S. Conference, to appear).

3. Conley, C.C. and R.W. Easton, Isolated invariant sets and isolating blocks, T.A.M.S. 158(1971).

4. Churchill, R.C., Isolated invariant sets in compact metric spaces, J. Diff. Eq. 13(1973), 523-550.

5. Montgomery, J.T., Cohomology of isolated invariant sets under perturbation, J. Diff. Eq. 13(1973), 257-299.

C. Conley
UNIVERSITY OF WISCONSIN

Joel Smoller
UNIVERSITY OF MICHIGAN

A TRANSITION FROM HOPF BIFURCATION TO CHAOS: COMPUTER EXPERIMENTS WITH MAPS ON R^2

by

James H. Curry* and James A. Yorke**

Numerical studies are used to investigate situations where $F: R^2 \to R^2$ and 0 is a fixed point which is repelling and F is dissipative in the large. That is, there is a compact attracting set C containing 0 such that for each x in R^2 there is an $n > 0$ such that that $F^n(x) \in C$ for all $n > N$. F is chosen to depend continuously on a parameter. We investigate situations where a curve appears which is seen (numerically) to be encircling 0 and to be attracting almost all trajectories. The parameter is changed continuously to cases where the attracting set is strange and chaotic. The transition cases are shown. When F is a homeomorphism, the strange chaotic attractors are one-dimensional.

§1. Introduction

The study of deterministic dynamical processes which have a somewhat chaotic behavior has a long history including large parts of the studies of statistical mechanics (and ergodic theory) and fluid dynamical turbulence. In recent years a number of authors have emphasized how easy it is for a dynamical process to be chaotic, and there has been a progression of ever simpler chaotic processes. Landau [1,2] argued that the irregularity of a turbulent fluid could come about by an infinite succession of bifurcations in the infinite dimensional phase space, each bifurcation adding a new oscillation frequency.

*The National Center for Atmospheric Research, sponsored by the National Science Foundation.

**Partially supported by National Science Foundation (MCS76-24432).

After n bifurcations, the attracting set would be an n dimensional torus T^n. He expected that as the stress parameter r (e.g., the Reynolds number) of the system was increased to some (apparently finite) critical value, r_c, n would go to ∞. Investigating a much lower dimensional situation, Ruelle and Takens [3] showed that after four bifurcations it would be possible to have a nonperiodic subset of the attracting torus T^4 be an attractor, and they coined the term "strange attractor" to describe the sets they found. Rather than investigate the ordinary differential equations, one frequently investigates surfaces transverse to the flow and the corresponding return map. In fact the principle example in [3] was presented in terms of a diffeomorphism of T^3 which has a strange attractor, and any orientation preserving (i.e., the Jacobian of the map is positive) diffeomorphism can be "suspended," that is, it is the return map for a flow on some space of one dimension higher.

The objective of this paper is to observe that chaos can be achieved via Hopf bifurcation in a manner quite different from that suggested in [3]. First our diffeomorphism in R^2 has an attracting invariant curve (topologically a circle) for some parameter value. As the parameter is changed the curve becomes distorted. As we change the parameter past a critical value the attracting set can develop folds, infinitely many, simultaneously, and this is what we observe in our pictures. The set is then chaotic and is no longer a curve. To distinguish between studies of chaotic behavior, we introduce the following concept.

We say a flow or map on R^n is <u>dynamically ergodic</u> if almost every point in R^n has the same positive limit set. This "almost" unique limit set can be a single point in simple cases. Bowen [4] also defines ergodic in a way that allows the limit set to be a small subset of the region of attraction. Notice that our perhaps unusual meaning of the word "ergodic" does not require that there exists an invariant measure on this limit set. (A related alternative approach might be to require the system to be "uniquely stable": all trajectories are bounded and there exists at least one compact Liapunov stable set, but there do not exist two disjoint compact Liapunov stable sets.) If the flow or map depends on a parameter we will say it is dynamically ergodic if it is dyamically ergodic for each parameter value in the interval of interest. The maps we examine here appear to be dynamically ergodic. In this sense our system is quite similar to those of Ruelle and Takens and Landau, and is quite different from the transition to chaos in the three-dimensional systems of Lorenz [5] and Curry [6].

The requirement of this uniqueness of the attracting set restricts the behavior somewhat in that additional attracting sets (possibly chaotic) cannot just pop up as the parameter is varied. In fact it seems hard to prove that any particular complex system is dynamically ergodic, yet numerical experiments often suggest that many systems studied are dynamically ergodic. We caution the reader that this dynamic ergodicity assumption does not require that the limit set (for almost every point) changes continuously in any sense as the parameter is varied.

Let us remark here the frequently investigated scalar dynamical process

$$x_{n+1} = rx_n(1-x_n) \qquad x_n \in [0,1],$$

appears to be dynamically ergodic. We restrict r to $[0,4]$ since for $r > 4$ almost every point seems to tend to $-\infty$. Julia [7] proved there could be at most one attracting periodic orbit. This simplest nonlinear dynamical process seems strangely similar to the infinite process described by Landau, starting with an attracting point for $0 \le r \le 1$, going through an infinite number of bifurcations as $r \to r_c$ $\approx 3.56998+$. At r_c there appears to be a stable Cantor set which is the positive limit set of almost every point in $[0,1]$. This Cantor set contains no periodic orbits and consists of almost periodic trajectories. (Notice that the Cantor set is a stable invariant set, and the smallest interval containing the Cantor set is also a stable invariant set, but there do not exist disjoint stable sets.) It is well known that for larger r there is non-almost-periodic behavior that in some ways is more chaotic than the infinite dimensional behavior described by Landau.

Chaotic phenomena occur in quite a variety of situations. See in particular the example of Rössler and his references. In most of these no study has been made of the transition from non-chaotic to chaotic behavior. The system in R^3 of Lorenz [4] has been investigated extensively [8-17], and it is not dynamically ergodic for $r < r_2 \approx 24.06$, the transition value mentioned by Lorenz. For $r \in (r_1, r_2) \approx (24.06, 24.74)$ there is a pair of stable equilibria and an attracting chaotic set. Landau and Ruelle and Takens, however, were investigating the transition to chaotic dynamics via systems that were or plausibly could be dynamically ergodic.

A number of investigations have been made of maps in the plane. Oxtoby and Ulam [18] showed that "most" continuous measure preserving maps of the plane are ergodic ("most" in the sense of Baire Category). The investigations of Moser [19] of sufficiently smooth area preserving twist maps on the annulus showed invariant curves could be expected, so in particular a general ergodic hypothesis was not valid for highly differentiable diffeomorphisms. Sitnikov [20] showed that in some situations there would exist chaotic oscillations and even a set which is homeomorphic to a shift. The non-area preserving maps are less well understood. Plis investigates [21,22] bounded invariant continua which do not separate the plane. The long-term behavior of iterates of maps in the plane has been investigated numerically in [25]. In particular, one of the simplest planar maps is investigated by Aronson and McGehee [27]. We feel our objective of investigating the transition to chaos via Hopf bifurcation is quite modest but is interesting and perhaps illuminating.

§2. Homeomorphisms

A homeomorphism on R^2 is a function which is continuous and has a continuous inverse. Our numerical investigations in this section concern mappings which are compositions of two simpler homeomorphisms. In polar coordinates define

$$\Psi_1(\rho,\Theta) = (\varepsilon \log(1+\rho),\ \Theta+\Theta_0)$$

where $\varepsilon \geq 1$ and $\Theta_0 \geq 0$ are parameters to be chosen.

Using Cartesian coordinates, define

$$\Psi_2(x,y) = (x, y+x^2)$$

and let Ψ denote the composition $\Psi_2 \circ \Psi_1$. Notice the mapping $\varepsilon \log(1+\rho)$ is a 1 to 1 function of ρ on $[0,\infty)$ and maps onto $[0,\infty)$ and has an inverse τ_ε. Hence $\Psi_1^{-1}(\rho,\Theta) = (\tau_\varepsilon(\rho),\ \Theta-\Theta_0)$ and $\Psi_2^{-1}(x,y) = (x,\ y-x^2)$ so Ψ is a homeomorphism. Consider first the behavior without Ψ_2. If we iterate Ψ_1 for $\varepsilon > 1$, all points except 0 tend asymptotically to an invariant circle (whose radius ρ_ε is the unique positive solution of $\rho_\varepsilon = \varepsilon \ell n(1+\rho_\varepsilon)$). As $\varepsilon \searrow 1$ the radius $\rho_\varepsilon \searrow 0$ so we may say a Hopf bifurcation occurs at $\varepsilon = 1$. For epsilon in the unit interval, $\Psi_1^n(p) \to 0$ as $n \to \infty$ for every $p \in R^2$. The mapping Ψ_2 introduces a nonlinearity which becomes more important as ε increases since the Jacobian of Ψ_2 at $(0,0)$ is the identity matrix. We have chosen the function $\varepsilon \ell n(1+\rho)$ in defining Ψ_1 since this function will dominate the polynomial nonlinearity in Ψ_2, in particular for any $n \geq 1$ and $\varepsilon \geq 1$, $[\varepsilon \ell n \rho]^n < \rho$ for large ρ.

For $\Theta_0 = 1$ and ε large an attracting fixed point dominates. The sequence we describe is for $\Theta_0 = 2$. Each display of the set consists of images of a single point. We chose a point, and iterated it one hundred times. Those initial iterates are not shown. Then we plotted the next four thousand iterates on the cathode ray tube.

The evolution of the attracting set. At $\varepsilon = 1$ the origin was observed to be a global attractor. The figures we display here are chosen from a large set. In cases where we see a connected set, it is possible that in reality the limit set has a large number of components packed closely together. If it is connected it is possible that a small change in ε would result in a much smaller figure. For example, at $\varepsilon = 1.40$ we see a connected set, but at $\varepsilon = 1.39$ (not displayed here) we observe an orbit with just three points.

At $\varepsilon = 1.01$ we see a nearly circular closed loop encircling the origin. As ε increases the size of the loop increases until $\varepsilon = 1.28$. From then until $\varepsilon_c = 1.3953+$ we found an orbit having period 3. These represent an attracting periodic orbit on an invariant curve which is homeomorphic (but not necessarily diffeomorphic) to a circle.

Immediately thereafter the connected loop returns, only it has wrinkles, an infinite set of them. At $\varepsilon = 1.42$ these wrinkles are small (see figure) and unimpressive. Nonetheless the figure is not homeomorphic to a circle. It is a strange attractor. By 1.45 (see figure) the wrinkles have grown. The bump containing P_1 in the upper right-hand corner is mapped smoothly to a sharpened cusp in the upper left, $\Psi(P_i) = P_{i+1}$, $i = 1,2,\ldots$. While the cusps at P_i appear sharp for $i > 1$, each is in fact the diffeomorphic image of its predecessor and so must be rounded. As we follow the sequence of cusps we see the P_4 cusp is showing signs of being flattened against the rest of the set. Each successive cusp is more flattened than the last. Notice, however, that the length of these cusps increases as long as we can perceive them. Soon the image of these cusps is squeezed so flat and long that we cannot follow them any longer. Yet this sequence of cusp points P_i is infinite. We are also seeing mixing: two points close to each other Q_1, Q_2 have images $\Psi^n(Q_1)$, $\Psi^n(Q_2)$ which are slowly stretched apart until they appear on opposite sides of P_1 and are folded closer together. This stretching and folding process mixes the images of Q_1 and Q_2 so that we may generally expect the pairwise images to separate and come closer together infinitely often.

At $\varepsilon = 1.52$ (see figure) we see these cusps growing faster and apparent mixing rate increased. This mixing rate is determined by the amount of the curve folded over.

At $\varepsilon = 1.63$ (see figure) the set has split into components. While this picture shows four thousand iterates, the next four thousand were found to produce essentially the same picture, no smaller, no larger. This picture is also virtually indistinguishable from that produced by plotting only the third or fourth thousand iterates. Each component in the picture is successively mapped onto the next and each component contains roughly one quarter of the total number of points.

At $\varepsilon = 1.70$, a connected figure is again seen. The region in the box is shown expanded in the next figure. Its boxed-in region enlarged in the next figure, and the increasingly detailed structures are reminiscent of the figures shown by Hénon.

At the transition value $\varepsilon_c = 1.3953...$, Ψ has an orbit of period 3. Then the eigenvalues of $D(\Psi^3)$ are 1.0 and approximately .47. (Note that $\theta_0 \approx 1/3$ of 2π.) We do not feel periodicity is crucial to the appearance of chaos. Rather, the basic agent seems to be that the nonlinearity creates a cusp which has infinitely many images, ever flatter and sharper.

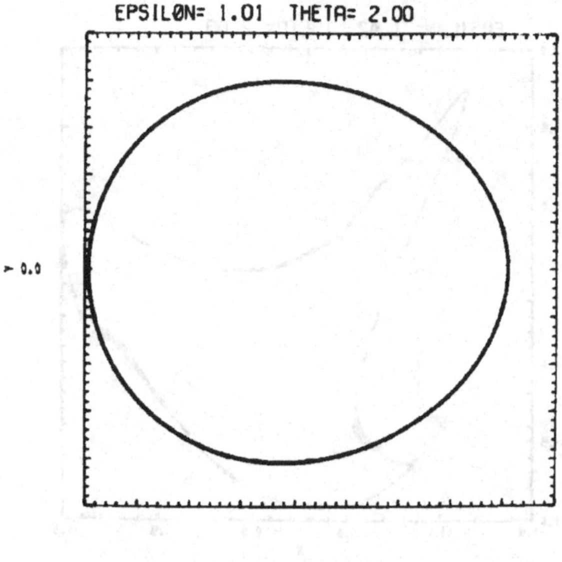
EPSILØN= 1.01 THETA= 2.00
Y 0.0
X

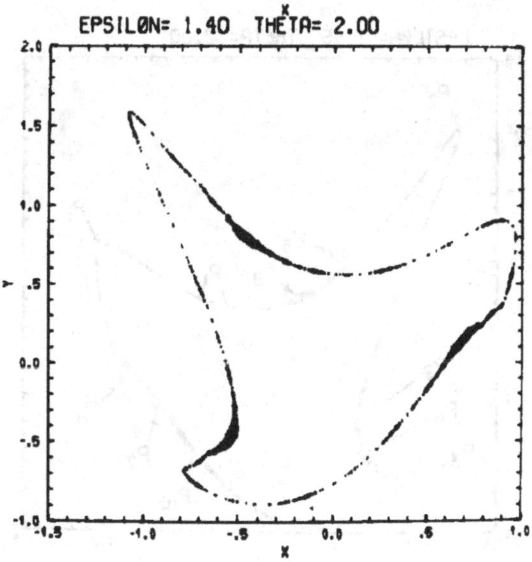
EPSILØN= 1.40 THETA= 2.00
Y
X

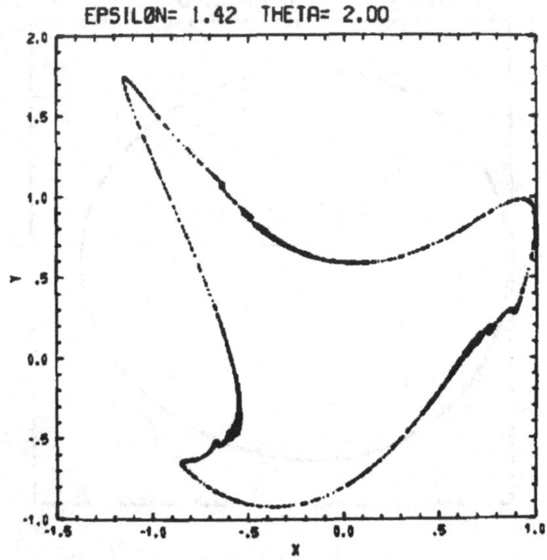
EPSILØN= 1.42 THETA= 2.00
Y
X

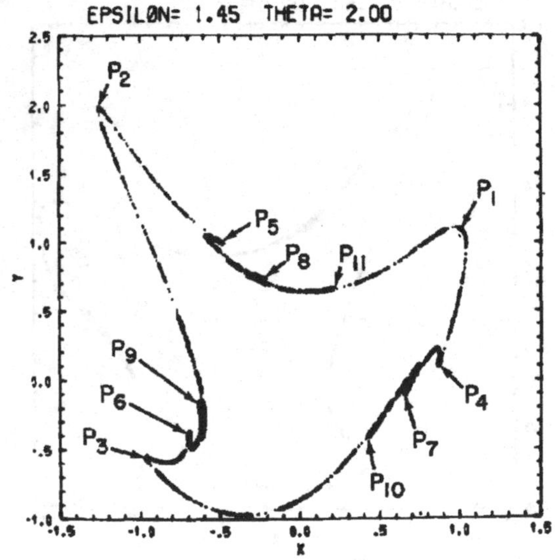
EPSILØN= 1.45 THETA= 2.00
Y
X
P1
P2
P3
P4
P5
P6
P7
P8
P9
P10
P11

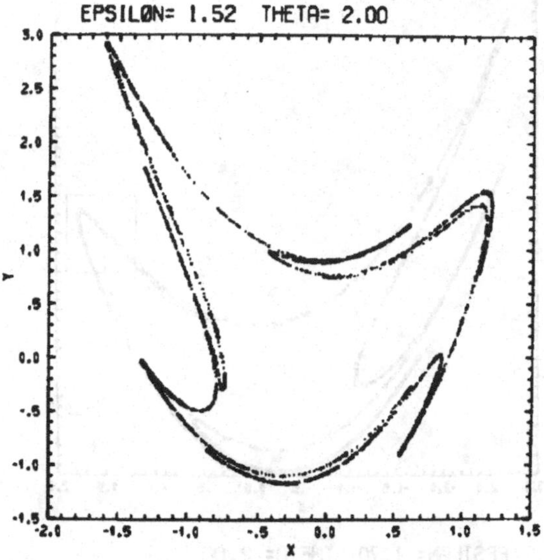
EPSILØN= 1.52 THETA= 2.00
Y
X

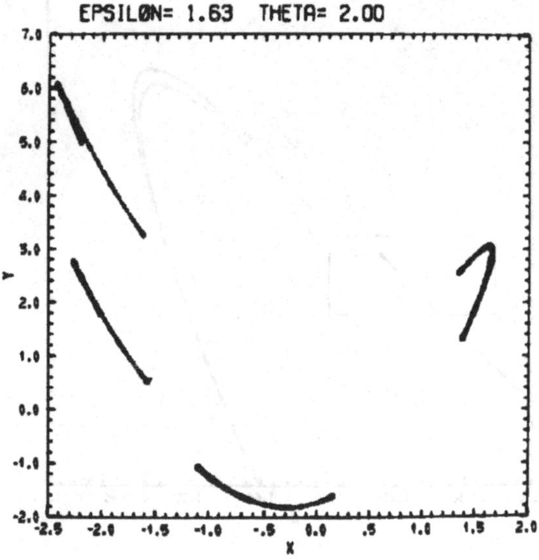
EPSILØN= 1.63 THETA= 2.00
Y
X

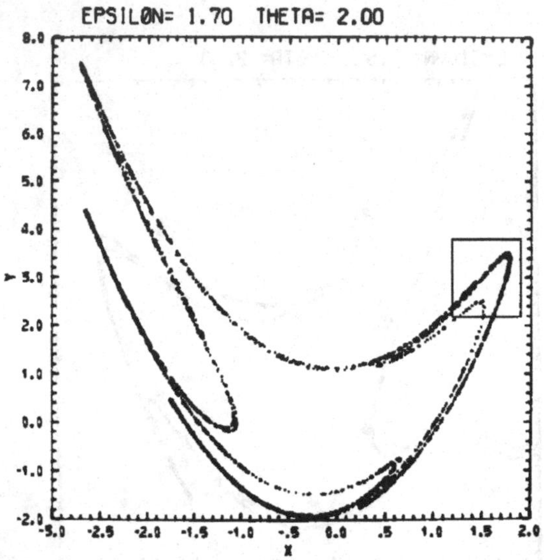
EPSILØN= 1.70 THETA= 2.00
8.0
7.0
6.0
5.0
4.0
3.0
2.0
1.0
0.0
-1.0
-2.0
Y
-3.0
-2.5
-2.0
-1.5
-1.0
-.5
0.0
.5
1.0
1.5
2.0
X

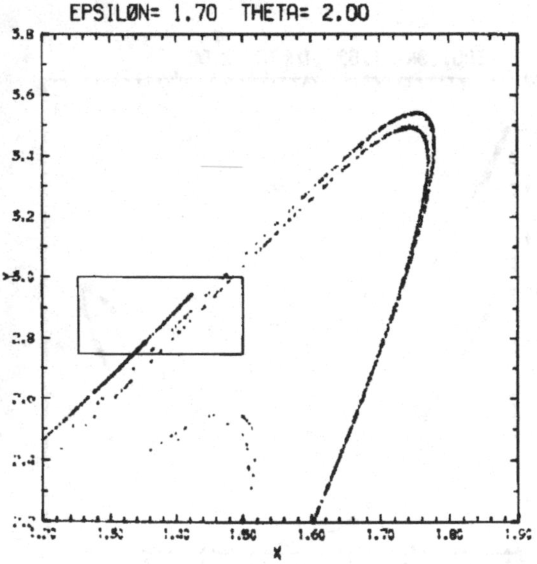
EPSILØN= 1.70 THETA= 2.00
3.8
3.6
3.4
3.2
Y
1.60
1.70
1.80
X

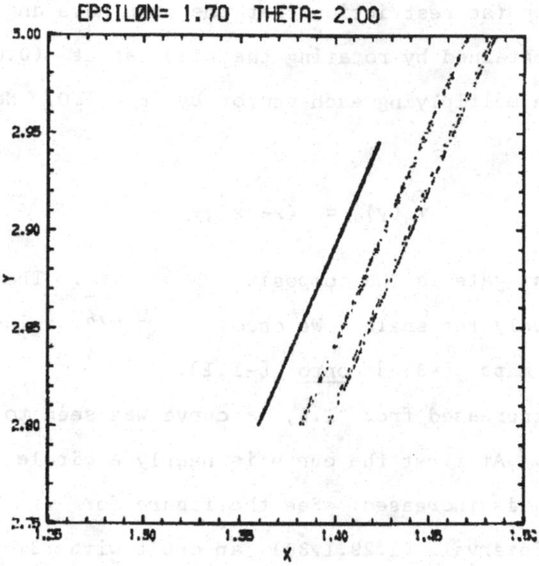
EPSILØN= 1.70 THETA= 2.00
3.00
2.95
2.90
2.85
2.80
2.75
Y
1.25
1.30
1.35
1.40
1.45
1.50
X

§3. A Non-Invertible Map

In this section we again look at a sequence of maps, though in this case we drop the restriction that the maps have an inverse. Let Ψ_1 be the map obtained by rotating the plane about (0,0) by one radian, and then multiplying each vector by $\varepsilon \geq 1.0$. Next define Ψ_2 by

$$\Psi_2(y) = (x-cx^3,y)$$

The map we investigate is the composite $\Psi = \Psi_2 \circ \Psi_1$. The constant $c > 0$ affects only the scale. We choose $c = 4/27$ since then the cubic $x - cx^3$ maps $[-3,3]$ <u>onto</u> $[-1,1]$.

As ε is increased from 1.0, a curve was seen to bifurcate from the origin. At first the curve is nearly a circle, but becomes irregular as ε is increased. See the figure for $\varepsilon = 1.28$. For epsilon in the interval (1.29,1.30) an orbit with 32 points is all that appears. As ε is increased slightly, a periodic orbit with period 58 appears. As ε is increased to 1.323 (see figure), the 58-point orbit has become unstable and is replaced by a set having 58 components. Careful examination shows this is a strange and chaotic set. As Ψ cyclically permutes these pieces we observe a stretching and folding analogous to that observed in the last section. Here however the mixing is limited: If Q_1 and Q_2 are chosen in the same component, then $\Psi^n(Q_1)$ will lie in the same component as $\Psi^n(Q_2)$.

As ε is increased, the 58 pieces grow until they merge (see figure for 1.33).

As ε increases we continue to see a set encircling the origin, and usually this region is connected. The figure for 1.28 appears possibly to be two-dimensional. In the figure for 1.35, the two-dimensionality becomes stronger. As ε is increased further, the open area inside the annulus shrinks until it is engulfed at 1.41 (see figure). When ε is increased further, we eventually reach a regime where there is no bounded invariant domain, and iterations of points usually escape to ∞.

In the last three figures for this section we iterated an initial point 25,000 times and displayed the final 24,000.

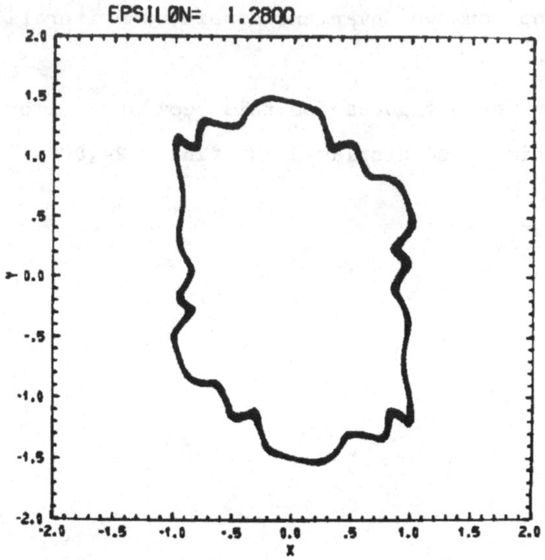
EPSILØN= 1.2800
Y
X

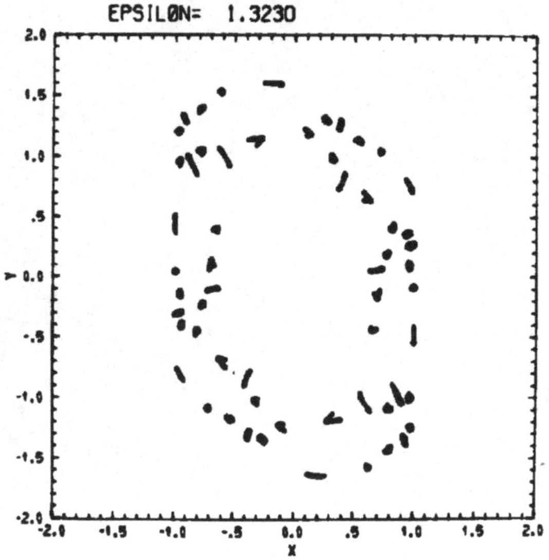
EPSILØN= 1.3230
Y
X

EPSILØN= 1.3300

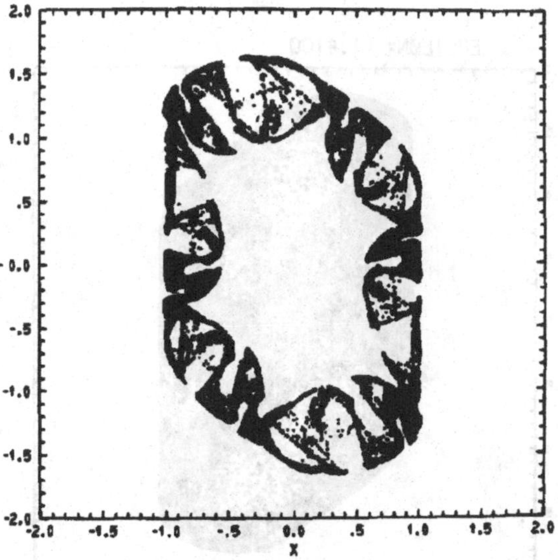

EPSILØN= 1.3500

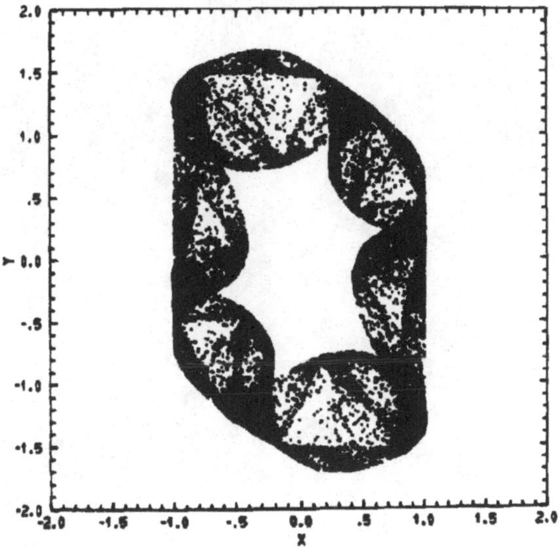

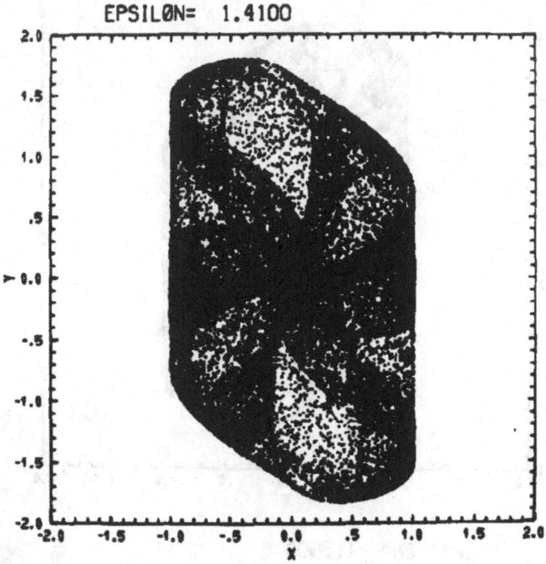
EPSILØN= 1.4100
2.0
1.5
1.0
.5
0.0
-.5
-1.0
-1.5
-2.0
Y
-2.0
-1.5
-1.0
-.5
0.0
.5
1.0
1.5
2.0
X

REFERENCES

1. L.D. Landau and E.M. Lifshitz, Fluid Mechanics, Pergamon Press, Oxford, 1959 (see pp. 105-107).

2. L.D. Landau, C.R. Acad. Sci., U.R.S.S. 44(1944), 311.

3. D. Ruelle and F. Takens, On the nature of turbulence, Comm. Math. Phys. 20(1971), 167-192; 23(1971), 343-344.

4. R. Bowen, A model for Couette flow data, in Springer Verlag Lecture Notes #615: Turbulence Seminar, 1977.

5. E.N. Lorenz, Deterministic nonperiodic flows, J. Atmos. Sci. 20 (1963), 130-141.

6. J. Curry, A generalized Lorenz system, preprint.

7. M.G. Julia, Mémoire sur l'itération des fonctions rationnelles, J. Math. Pures et Appl., Serie 7 tome 4(1918), 47-245.

8. J.L. Kaplan and J.A. Yorke, Preturbulence: a regime observed in a fluid flow model of Lorenz, a preprint.

9. J.L. Kaplan and J.A. Yorke, The onset of chaos in a fluid flow model of Lorenz, in Proceedings of the New York Acad. of Sci. meeting on Bifurcation, held in November 1977 in New York City.

10. J.A. Yorke and E.D. Yorke, Metastable chaos: the transition to sustained chaotic oscillations in a model of Lorenz, a preprint.

11. K.A. Robbins, A new approach to subcritical instability and Turbulent transitions in a simple dynamo, Math. Proc. Cambridge Phil. Soc., to appear.

12. O.E. Rossler, Horseshoe-map chaos in the Lorenz equation, Physics Letters 60A(1977), to appear.

13. Efraimovich, Bikov, and Silnikov, The origin and structure of the Lorenz attractor, Dokl. Acad. Nauk SSR 234(1977), 336-339.

14. M. Lucke, Statistical dynamics of the Lorenz model, J. Statistical Physics 15(1976), 455-474.

15. J. Guckenheimer and R.F. Williams, to appear.

16. J.B. McLaughlin and P.C. Martin, Transition to turbulence in a statically stressed fluid system, Phys. Rev. A12(1975), 186-203.

17. O.E. Lanford, Qualitative and statistical theory of dissipative systems (preprint).

18. J.C. Oxtoby and S.M. Ulam, Measure preserving homeomorphisms and metrical transitivity, Ann. Math. 42(1941), 87-92.

19. J. Moser, A rapidly convergent iteration method, Part II, Ann. Scuola Norm. Sup. Pisa 20(1965), 499-535.

20. K. Sitnikov, Existence of oscillating motions for the three-body problem, Dokl. Akad. Nauk 133(1960), 303-306.

21. V.A. Plis, On recurrent motions in periodic systems of two differential equations, Differentsial'nye Uravneniya 3(1967), 722-732.

22. V.A. Plis, Some problems in the behavior of solutions of periodic dissipative second-order systems, ibid, 2(1966).

23. F. Rannou, Numerical study of discrete plane area-preserving mappings, Astron. and Astrophys. 31(1974), 289-301.

24. H. Hénon, A two-dimensional mapping with a strange attraction, Comm. Math. Phys. 50(1976), 69-77.

25. J.R. Beddington, C.A. Free, and J.H. Lauton, Dynamic complexity in predator-prey models framed in difference equations, Nature 255(1975), 58-60.

26. P. Stein and S. Ulam, Nonlinear transformation studies on electronic computers, Rozprawy Metamat. 39(1964), 401-484.

27. D. Aronson and R. McGehee, in this volume.

James H. Curry
HOWARD UNIVERSITY
and
THE NATIONAL CENTER FOR ATMOSPHERIC RESEARCH

James A. Yorke
UNIVERSITY OF MARYLAND

TRANSVERSE HETEROCLINIC ORBITS
IN THE ANISOTROPIC KEPLER PROBLEM

by

Robert L. Devaney*

§1. Introduction

The Anisotropic Kepler Problem is a one parameter family of classical mechanical systems recently introduced by Gutzwiller to approximate certain quantum mechanical systems. In particular, this system arises naturally when one looks for bound states of an electron near a donor impurity of a semi-conductor. Here the potential is due to an ordinary Coulomb field, while the kinetic energy becomes anisotropic because of the electronic band structure in the solid. Gutzwiller [9] suggests that this situation is akin to an electron whose mass in one direction is larger than in the other directions.

Aside from these physical implications, the Anisotropic Kepler problem also exhibits many phenomena of considerable mathematical interest; we deal exclusively with this aspect of the problem in the sequel. For more details of the physical applications of this work, we refer to [8].

When the parameter $\mu = 1$, the problem reduces to the ordinary Kepler or central force problem. If we choose coordinates $\underline{q} = (q_1, q_2)$ in the configuration space R^2 together with momentum coordinates $\underline{p} = (p_1, p_2)$, then the orbits of the system are given by the flow of the vector field

$$\begin{aligned} \dot{\underline{q}} &= \underline{p} \\ \dot{\underline{p}} &= -\underline{q}/|q|^3 \end{aligned} \tag{1.1}$$

on R^4. This system is Hamiltonian with Hamiltonian function

*Partially supported by National Science Foundation (MPS74-06731).

$$H(\underline{q},\underline{p}) = \frac{1}{2}\,\underline{p}^{\,2} - 1/|\underline{q}|. \tag{1.2}$$

When the parameter $\mu > 1$, the kinitic energy of the system becomes anisotropic. Let M be the 2×2 matrix

$$\begin{bmatrix} \mu & 0 \\ 0 & 1 \end{bmatrix}. \tag{1.3}$$

Then the new Hamiltonian or total energy function is

$$H(\underline{p},\underline{q}) = \frac{1}{2}\underline{p}^t M \underline{p} - 1/|\underline{q}|. \tag{1.4}$$

Via Hamilton's equations, the system is thus given by the vector field

$$\begin{aligned} \dot{\underline{q}} &= M\underline{p} \\ \dot{\underline{p}} &= -\underline{q}/|\underline{q}|^3. \end{aligned} \tag{1.5}$$

Of course, H is an integral for this system; that is, orbits of (1.5) are constrained to lie on the invariant level sets or energy surfaces $H = e$. We henceforth consider only negative energy levels.

As we shall see below, the introduction of anisotropy into the kinetic energy changes the orbit structure of the Kepler problem dramatically. For one thing, the nature of the singularity at $\underline{q} = \underline{0}$ changes as μ increases. When $\mu = 1$, the singularity is regularizable by any of several methods [6,14]. This means that orbits of (1.5) which meet $\underline{q} = \underline{0}$ in either time direction can be continued in some "reasonable" fashion through $\underline{q} = \underline{0}$. This is no longer true for most $\mu > 1$. In these cases, nearby initial conditions give rise to orbits which leave a neighborhood of $\underline{q} = \underline{0}$ in opposite directions. There is no way that orbits which approach $\underline{q} = \underline{0}$ can be connected to orbits leaving $\underline{q} = \underline{0}$ in a globally continuous fashion. We touch on the regularization question in Section 3. For more details, however, we refer to [5].

A second point of departure from the Kepler problem is the

non-integrability of the Anisotropic Kepler problem. When $\mu = 1$, the Kepler problem is completely integrable; that is, the angular momentum $\underline{q} \times \underline{p}$ is a constant of the motion which is independent of the total energy. For most $\mu > 9/8$, there is no such second independent integral. This is proved in [4].

The reason for the non-integrability of the Anisotropic Kepler problem is the existence of quite a bit of "pathology" or randomness in the system. When $\mu = 1$, all orbits of the system are either closed and lie on ellipses encircling $\underline{q} = \underline{0}$, or else are singular and lie on a cylinder of orbits which both begin and end in collision with $\underline{q} = \underline{0}$. These are the only possibilities for $\mu = 1$.

For $\mu > 1$, however, this simple phase portrait is destroyed. As μ increases, orbits tend to oscillate more and more wildly about the q_2-axis. By the time μ reaches $9/8$, one can find isolated orbits which cross the q_2-axis an arbitrarily large number of times before closing up. One can also find isolated collision orbits which also oscillate about the q_2-axis before reaching $\underline{q} = \underline{0}$. We show the existence of both types of orbits in Section 4 below. But, in fact, one can show much more. There is a subsystem of the Anisotropic Kepler problem which is topologically conjugate to a Bernoulli shift on infinitely many symbols. This means that, in addition to arbitrarily long closed orbits, there exist non-periodic orbits which come arbitrarily close to any of the closed orbits in the subsystem. This fact is shown in [4]. We remark that the orbits of Section 4 form only a small part of this subsystem.

The reason for this complicated orbit structure is the existence of a cycle of transverse heteroclinic solutions for the Anisotropic Kepler problem. Actually, these heteroclinic solutions are orbits which begin and end at the singularity, but which are viewed in a different time scale. Often, finding such heteroclinic solutions in a given dynamical system is relatively easy, but proving that they are transverse is usually more difficult. Our method makes use of techniques due to Conley and is presented in detail in Section 3.

§2. The Collision Manifold

Our goal in this section is to study how orbits behave as they pass close to the singularity at $\underline{q} = \underline{0}$. To accomplish this, we introduce an invariant manifold at "collision" over which the vector field extends analytically. The behavior of orbits near collision is then governed by the flow on this manifold. This technique has been exploited by McGehee [11] in his study of triple collision in the collinear three body problem.

We first introduce new variables via

$$\begin{aligned} \underline{q} &= r\underline{s} \\ \underline{p} &= r^{-1/2}\underline{u} \end{aligned} \tag{2.1}$$

where $r \in (0,\infty)$ and $\underline{s}$ is a point on the unit circle S^1. The differential equation becomes

$$\begin{aligned} \dot{r} &= r^{-1/2}\underline{s}^t M\underline{u} \\ \dot{\underline{s}} &= r^{-3/2}(M\underline{u} - (\underline{s}^t M\underline{u})\underline{s}) \\ \dot{\underline{u}} &= r^{-3/2}(\tfrac{1}{2}(\underline{s}^t M\underline{u})\underline{u} - \underline{s}) \end{aligned} \tag{2.2}$$

and the total energy relation becomes

$$1 + re = \frac{1}{2}\underline{u}^t M\underline{u}. \tag{2.3}$$

Here e is the constant value of the total energy, which we assume to be negative. The singularity at $\underline{q} = \underline{0}$ now corresponds to the boundary $r = 0$ of the open manifold $(0,\infty) \times S^1 \times R^2$.

We now analytically extend the vector field defined by (2.2) over this boundary. This is accomplished by the change of time scale

$$dt = r^{3/2}\, d\xi. \tag{2.4}$$

In the new time scale, the system (2.2) goes over to

$$\begin{aligned} \dot{r} &= r(\underline{s}^t M\underline{u}) \\ \dot{\underline{s}} &= M\underline{u} - (\underline{s}^t M\underline{u})\underline{s} \\ \dot{\underline{u}} &= \tfrac{1}{2}(\underline{s}^t M\underline{u})\underline{u} - \underline{s}. \end{aligned} \tag{2.5}$$

This is an analytic vector field on $[0,\infty) \times S^1 \times R^2$. Note that $r = 0$ is an invariant manifold for the flow. We have replaced the singularity at $\underline{q} = \underline{0}$ by a "collision manifold." Orbits which previously reached the singularity (in either time direction) are now slowed down and tend asymptotically to the collision manifold. And orbits which previously passed close to collision now behave very much like orbits on the collision manifold itself. We discuss this subflow next.

On $r = 0$, the energy relation (2.3) gives

$$\tfrac{1}{2}\underline{u}^t M\underline{u} = 1. \tag{2.6}$$

This defines an invariant torus in $r = 0$ which we denote by Λ. As (2.6) is independent of the total energy, it follows that Λ forms the invariant boundary of each energy surface in phase space.

Using (2.5), it follows that the flow on Λ is given by

$$\begin{aligned} \dot{\underline{s}} &= M\underline{u} - (\underline{s}^t M\underline{u})\underline{s} \\ \dot{\underline{u}} &= \tfrac{1}{2}(\underline{s}^t M\underline{u}) - \underline{s}. \end{aligned} \tag{2.7}$$

To study this flow, it is convenient to introduce new variables

$$\begin{aligned} \underline{s} &= (\cos\theta, \sin\theta) \\ \underline{u} &= \sqrt{2(1+re)}\,(\cos\psi, \sin\psi) \\ d\xi &= \sqrt{2(1+re)}\, d\tau. \end{aligned} \tag{2.8}$$

The differential equation (2.5) is transformed into

$$\dot{r} = 2(1+re)(r)(\mu^{1/2}\cos(\psi)\cos(\theta) + \sin(\psi)\sin(\theta))$$

$$\dot{\theta} = 2(1+re)(\sin(\psi)\cos(\theta) - \mu^{1/2}\cos(\psi)\sin(\theta)) \qquad (2.9)$$

$$\dot{\psi} = \mu^{1/2}\sin(\psi)\cos(\theta) - \cos(\psi)\sin(\theta)$$

where the dot indicates differentiation with respect to τ. We denote this vector field by X_μ. Restricted to Λ, the system becomes

$$\dot{\theta} = 2(\sin(\psi)\cos(\theta) - \mu^{1/2}\cos(\psi)\sin(\theta))$$
$$\dot{\psi} = \mu^{1/2}(\sin(\psi)\cos(\theta) - \cos(\psi)\sin(\theta)) . \qquad (2.10)$$

It is this flow which governs the behavior of orbits near collision.

At this point we remark that, for negative energy, the vector field (2.9) is an analytic system on $[0,-1/e] \times T^2$, where T^2 denotes the two dimensional torus $S^1 \times S^1$. The boundary $\{0\} \times T^2$ is the collision manifold which we study in this section; the boundary $\{-1/e\} \times T^2$ corresponds to the "oval of zero velocity" in the original system. The flow on this component of the boundary is important for the transversality of the heteroclinic orbits and will be discussed in the next section.

Returning to the flow on Λ, one may check easily that, for $\mu > 1$, there are exactly eight equilibrium solutions for the system (2.10). Their locations as well as their characteristic exponents are given in Table I. We remark that Table I gives both the two exponents in directions tangent to Λ as well as the exponent in the "normal" direction. We also note that each equilibrium point is hyperbolic, and that, for the restriction of X_μ to Λ, there are two sinks, two sources, and four hyperbolic saddle points.

We note several other important qualitative features of the flow on Λ:

1. If $\mu > 9/8$, then $\sqrt{9-8\mu}$ is imaginary. Hence the characteristic exponents in directions tangent to Λ at both the sinks

and the sources are complex, as we see from Table I. This means that nearby orbits in Λ tend to spiral into and away from the corresponding sinks and sources. This phenomenon is known to be necessary for the existence of Bernoulli shifts [3], and is the reason why μ must be chosen greater than 9/8.

2. For all $\mu > 1$, there are no other non-wandering points for the restriction of X_μ to Λ. Indeed, the real-valued function defined by

$$f_\mu(\underline{s},\underline{u}) = |M^{-1/2}\underline{s}|^{-1/2}(\underline{s}^t\underline{u})$$

where $M^{-1/2}$ is the 2×2 matrix

$$M^{-1/2} = \begin{bmatrix} \mu^{-1/2} & 0 \\ 0 & 1 \end{bmatrix}$$

increases along all non-equilibrium orbits. This means that the flow on Λ is gradient-like and, consequently has no closed or recurrent orbits.

3. Each of the four saddle points in Λ admits one dimensional stable and unstable manifolds. As a consequence of 2) above, there are two possibilities for the ultimate behavior of each of these curves. Either each branch of all the unstable manifolds is forward asymptotic to one of the sinks, or else one or more of them match up with a stable manifold of another saddle point (that is, we have a saddle connection). That this second possibility does not occur for an open and dense set of $\mu > 1$ is proven in [4]. Thus we have that for almost all $\mu > 1$, all stable and unstable manifolds of the saddle points are ultimately asymptotic to sources and sinks.

4. Let p be one of the saddles in Λ. For most $\mu > 1$, both

branches of the unstable manifold $W^u(p)$ are forward asymptotic to one of the two sinks. We claim that each branch dies in a distinct sink. Indeed, the map $(\theta,\psi) \to (-\theta,-\psi)$ preserves the system (2.10). This map fixes each saddle point in Λ, but interchanges the two sinks. Also, the two branches of $W^u(p)$ are interchanged. Hence, elementary symmetry arguments yield the result. Similarly, each branch of the stable manifold $W^s(p)$ emanates from distinct sources. This fact is important for the non-regularization result mentioned above. See [5].

TABLE 1

Equilibrium point	Characteristic Exponents on Λ	off Λ	Type on Λ
$(-\frac{\pi}{2}, -\frac{\pi}{2})$	$-\frac{1}{2} \pm \frac{1}{2}\sqrt{9-8\mu}$	2	Sink
$(0,0)$	$-\frac{\sqrt{\mu}}{2} \pm \frac{1}{2}\sqrt{9\mu-8}$	$2\sqrt{\mu}$	Saddle
$(\frac{\pi}{2}, \frac{\pi}{2})$	$-\frac{1}{2} \pm \frac{1}{2}\sqrt{9-8\mu}$	2	Sink
(π,π)	$-\frac{\sqrt{\mu}}{2} \pm \frac{1}{2}\sqrt{9\mu-8}$	$2\sqrt{\mu}$	Saddle
$(-\frac{\pi}{2}, \frac{\pi}{2})$	$\frac{1}{2} \pm \frac{1}{2}\sqrt{9-8\mu}$	-2	Source
$(0,\pi)$	$\frac{\sqrt{\mu}}{2} \pm \frac{1}{2}\sqrt{9\mu-8}$	$-2\sqrt{\mu}$	Saddle
$(\frac{\pi}{2}, -\frac{\pi}{2})$	$\frac{1}{2} \pm \frac{1}{2}\sqrt{9-8\mu}$	-2	Source
$(\pi,0)$	$\frac{\sqrt{\mu}}{2} \pm \frac{1}{2}\sqrt{9\mu-8}$	$-2\sqrt{\mu}$	Saddle

Figure 1 below gives a sketch of the phase portrait of X_μ on Λ when $\mu > 9/8$. We sketch only the case where there are no saddle connections.

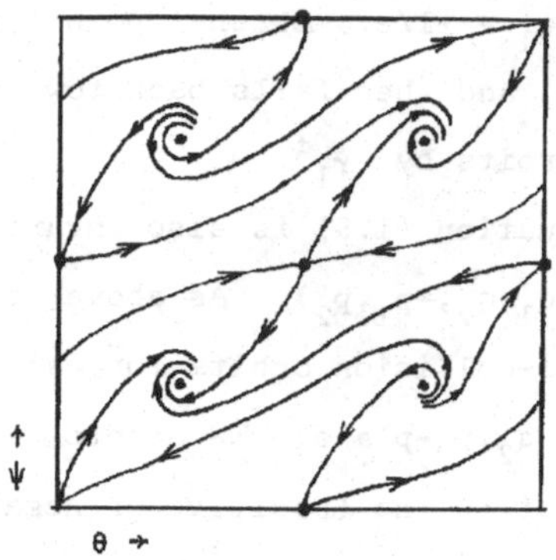

Figure 1. The flow on the collision manifold.

§3. Transverse Collision Orbits

We turn now to the main result of this paper. An orbit of the Anisotropic Kepler problem is called a _bi-collision orbit_ if it is singular in both time directions, i.e., the orbit meets $\underline{q} = \underline{0}$ at two different times. For the Kepler problem, all of these orbits are well understood. For negative total energy, there is a one parameter family of such orbits. Each lies along a ray θ = constant in configuration space and has a unique point of zero velocity. When $\mu > 1$, most of these bi-collision orbits are destroyed. Some, however, persist. In particular, there are four _primary bi-collision_ orbits which we now define.

The original differential equation is invariant under the reflection $(q_1,q_2,p_1,p_2) \to (q_1,-q_2,p_1,-p_2)$. Hence orbits passing through points of the form $(q_1,0,p_1,0)$ are trapped in the q_1,p_1-plane; they project to orbits which travel along the q_1-axis in configuration space. Furthermore, for negative energy, (1.4) implies that such orbits lie within the circle $r = -1/e$ in the q_1,q_2-plane. Hence it follows easily that, for each negative energy level, there are exactly two bi-collision orbits trapped on the q_1-axis. Each leaves $\underline{q} = \underline{0}$ with infinite velocity and travels along either the positive or negative

q_1-axis until reaching $q_1 = \pm 1/e$. At that point, the particle momentarily has zero velocity, and then falls back toward $\underline{q} = \underline{0}$. We denote these two bi-collision orbits by $\gamma_1^{\pm}$.

The differential equation (1.5) is also invariant under the reflection $(q_1,q_2,p_1,p_2) \to (-q_1,q_2,-p_1,p_2)$. As above, this forces the existence of two additional bi-collision orbits for each negative energy level, this time in the q_2,p_2-plane. We denote these orbits by $\gamma_2^{\pm}$ and note that they project to the (positive or negative) q_2-axis in configuration space.

The change of time scale (2.4) has the effect of slowing the primary bi-collision orbits down so that they approach Λ asymptotically in both time directions. Since the flow on Λ is gradient-like, this implies that each such orbit is asymptotic to an equilibrium point in Λ. Which equilibrium point these bi-collision orbits approach is obvious from the change of variables (2.8). We simply summarize this data as follows:

Proposition. *Let* $W^s(p)$ *and* $W^u(p)$ *denote the stable and unstable manifolds at the equilibrium point* p. *Then*

i. $\gamma_1^{+} \subset W^s(0,\pi) \cap W^u(0,0)$

ii. $\gamma_1^{-} \subset W^s(\pi,0) \cap W^u(\pi,\pi)$

iii. $\gamma_2^{+} = W^s(\pi/2, -\pi/2) \cap W^u(\pi/2,\pi/2)$

iv. $\gamma_2^{-} = W^s(-\pi/2,\pi/2) \cap W^u(-\pi/2,\pi/2)$.

Again using Table I, the dimensions of $W^s(\pm\pi/2, \mp\pi/2)$ and $W^u(\pm\pi/2, \pm\pi/2)$ are all one; this is the reason for the equality in iii and iv above. In contrast, the dimensions of the remaining stable and unstable manifolds are all two. Hence it is natural to ask whether or not these invariant manifolds meet transversely (within the three-dimensional energy surfaces) along $\gamma_1^{\pm}$. Our result is that this is indeed the case for all $\mu > 1$.

The basic idea of the proof is to watch how $W^s(0,\pi)$ and $W^u(0,0)$ meet near the so-called oval of zero velocity. This is the circle given by $r = -1/e$, $\underline{p} = \underline{0}$ in each energy surface. Orbits of points on this circle are symmetric in the sense that their projections to configuration space in both forward and backward time are identical curves. We shall make use of this symmetry in the next section. We denote by $Z = Z_e$ the oval of zero velocity in the energy level $H^{-1}(e)$.

To study how $W^s(0,\pi)$ and $W^u(0,0)$ behave near Z, we blow up this circle into a torus. This is also accomplished by the change of variables (2.8). The resulting system (2.9) is defined on $[0,-1/e]\times T^2$, with both boundaries $\{0\}\times T^2$ and $\{-1/e\}\times T^2$ invariant under the flow. The component $\{0\}\times T^2$ corresponds to Λ, while $\{-1/e\}\times T^2$ corresponds to the oval of zero velocity. We denote this component of the boundary by Ω. Orbits which previously crossed Z are "broken" into two pieces: one orbit prior to the crossing of Z which now tends asymptotically to Ω, and a post-crossiing orbit which tends toward Ω in the backward time direction. That is, orbits which previously crossed Z now lie in either the stable or the unstable manifold of Ω, which we denote by $W^s(\Omega)$ and $W^u(\Omega)$. Orbits which previously came close to Z can now be studied relative to these invariant sets.

Restricted to Ω, the system (2.9) is given by

$$\begin{aligned} \dot{\theta} &= 0 \\ \dot{\psi} &= \mu^{1/2}\sin(\psi)\cos(\theta) - \cos(\psi)\sin(\theta) . \end{aligned} \qquad (3.1)$$

This system is easily solved: for all μ there are two circles of equilibria in Ω given by

$$\mu^{1/2}\sin(\psi)\cos(\theta) = \cos(\psi)\sin(\theta) . \qquad (3.2)$$

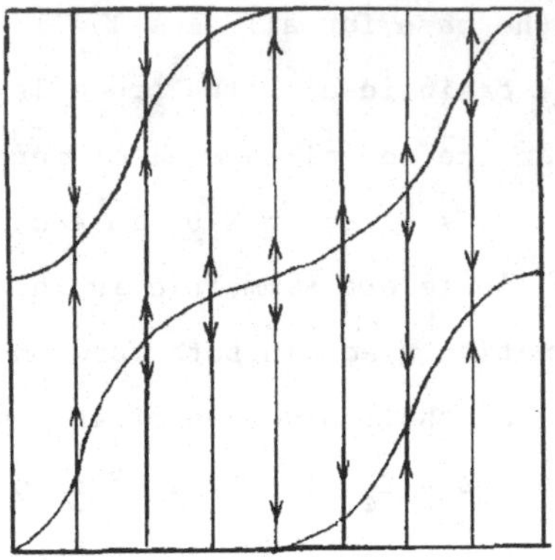

Figure 2. The flow on Ω.

See Figure 2. We let C_1 be the circle of equilibria passing through the point $(-1/e,0,0)$; let C_2 be the other circle.

One computes that both circles are normally hyperbolic with two dimensional stable and unstable manifolds. Also, C_1 is a repellor for the restriction of the flow to Ω, and C_2 is an attractor.

We observe that both C_1 and C_2 meet the circle $\theta = 0$ in Ω at an angle $\beta(\mu)$ where, for $\mu > 1$, $\beta(\mu)$ satisfies

$$\pi/4 < \beta(\mu) < \pi/2. \tag{3.3}$$

This can be seen by implicit differentiation of (3.2).

We now proceed to the proof of the

Theorem. <u>For all</u> $\mu > 1$, $W^s(0,\pi)$ <u>meets</u> $W^u(0,0)$ <u>transversely</u> (<u>within each negative energy surface</u>) <u>along the primary bi-collision orbit</u> γ_1^+. <u>Also</u>, $W^s(\pi,0)$ <u>meets</u> $W^u(\pi,\pi)$ <u>transversely along</u> γ_1^-.

<u>Proof</u>: We prove that $W^s(0,\pi)$ meets $W^u(0,0)$ transversely; the other case is similar.

Let $T = T(r_0)$ denote the torus given by $r = r_0$ in $[0,-1/e] \times T^2$. If r_0 is close enough to $-1/e$, then T is an isolating block for the invariant set Ω in the sense of [1]. For each r_0, $0 \le r_0 < -1/e$, $W^s(0,\pi)$ meets T transversely at the point $(r_0,0,\pi)$. Hence there is a smooth curve $\gamma_s(\mu,r_0)$ passing through $(r_0,0,\pi)$ in $T \cap W(0,\pi)$. Let $\alpha_s(\mu,r_0)$ denote the angle this curve makes with $\theta = 0$ in T. The lemma below shows that

$$0 < \alpha_s(\mu,r_0) < \pi/4 \tag{3.4}$$

if $\mu > 1$ and $0 \le r_0 < -1/e$.

Assuming this result, the proof is completed as follows. For each r_0, $0 < r_0 < -1/e$, $W^u(\Omega)$ also meets T transversely at $(r_0,0,\pi)$. Denote the angle of intersection of $W^u(\Omega) \cap T(r_0)$ with $\theta = 0$ by $\beta_u(\mu,r_0)$. By (3.3) it follows that

$$\pi/4 < \beta_u(\mu,r_0) < \pi/2 \tag{3.5}$$

for r_0 close enough to $-1/e$.

Now the system (2.9) is reversed by the symmetry

$$(r_0,\theta,\psi) \to (r_0,\theta,-\psi). \tag{3.6}$$

In particular, $W^s(\Omega)$ is mapped to $W^u(\Omega)$ and $W^s(0,\pi)$ to $W^u(0,0)$ by this mapping. Hence we have similar results near the point $(r_0,0,0)$ for $W^u(0,0) \cap T(r_0)$ and $W^s(\Omega) \cap T(r_0)$. Let $\alpha_u(\mu,r_0)$ and $\beta_s(\mu,r_0)$ denote the angles of intersection of these curves with $\theta = 0$ at $(r_0,0,0)$. We have

$$\begin{aligned} 0 &< \alpha_u(\mu,r_0) < \pi/4 \\ \pi/4 &< \beta_s(\mu,r_0) < \pi/2 \end{aligned} \tag{3.7}$$

as above. Figure 3 shows the relative positions of these various curves in T.

Now one computes easily that

$$\dot{\theta} = 2(1+re)\sin(\psi) \tag{3.8}$$

along $\theta = 0$. Hence $\dot{\theta} > 0$ for $\theta = 0$, $0 < \psi < \pi$.

Let $\gamma(s)$ be a small arc in $W^u(0,0) \cap T(r_0)$ passing through $\gamma(0) = (r_0,0,0)$. Suppose $\theta(\gamma(s)) > 0$ for $s > 0$. The forward orbits of all points $\gamma(s)$ with $s \neq 0$ eventually reintersect $r = r_0$. Let $\hat{\gamma}(s)$ denote the first such reintersection, for $s \neq 0$. Then $\hat{\gamma}(s)$ is a smooth curve in $T(r_0)$ satisfying

$$\lim_{s \to 0} \hat{\gamma}(s) = (r_0, 0, \pi) = \hat{\gamma}(0).$$

In fact, $\hat{\gamma}$ is smooth even at $s = 0$ since, in terms of the original system, $\hat{\gamma}$ may be obtained from γ by applying an ordinary Poincare map along the orbit γ_1^+.

Now (3.8) implies that $\theta(\hat{\gamma}(s)) > 0$ for $s > 0$. Moreover, using the flow of (3.1), it follows immediately that $\hat{\gamma}(s)$ is contained for small s in the sector bounded by $\theta = 0$ and $W^u(\Omega)$. This implies that $W^s(0,\pi)$ and $W^u(0,0)$ are not tangent at $(r_0, 0, 0)$, and hence that they are never tangent. This completes the proof with the exception of the lemma.

Lemma. <u>If</u> $\mu > 1$ <u>and</u> $0 \leq r_0 < -1/e$, <u>then</u> $0 < \alpha_s(\mu, r_0) < \pi/4$.

<u>Proof</u>: The idea here is to construct a Wazewski set for the flow as in [2]. Using (2.10), one computes easily that the stable eigenspace at $(0,0,\pi)$ in Λ is given by the line

$$\psi' = \frac{1}{4}(3\mu^{1/2} + \sqrt{9\mu - 8})\,\theta'$$

in the tangent space to Λ. The slope of this line is greater than one, and so the result is true for $r_0 = 0$.

Now consider the submanifolds $\theta = 0$ and $\psi = \theta + \pi$ near γ_1^+. Let D be the sector satisfying

$$0 \leq \theta \leq \psi - \pi$$
$$\pi < \psi < 3\pi/2.$$

Along $\theta = 0$ we have

$$\dot{r} \leq 0$$
$$\dot{\theta} = 2(1+re)\sin(\psi)$$
$$\dot{\psi} = \mu^{1/2}\sin(\psi).$$

Hence both $\dot{\theta}$ and $\dot{\psi}$ are negative along $\theta = 0$, for $\pi < \psi < 2\pi$.

On the other hand, along $\theta = \psi - \pi$, we have

$$\dot{r} \leq 0$$

$$\dot{\theta} = (1+re)(\mu^{1/2}-1)\sin(2\psi)$$

$$\dot{\psi} = \frac{1}{2}(1-\mu^{1/2})\sin(2\psi).$$

Hence $\dot{\theta} > 0$ and $\dot{\psi} < 0$ for $\pi < \psi < 3\pi/2$. Thus orbits tend to leave the sector D in forward time, at least near γ_1^+. See Figure 4. It follows that $W^s(0,\pi)$ is trapped (at least locally) in the sector D together with its reflection about γ_1^+. qed

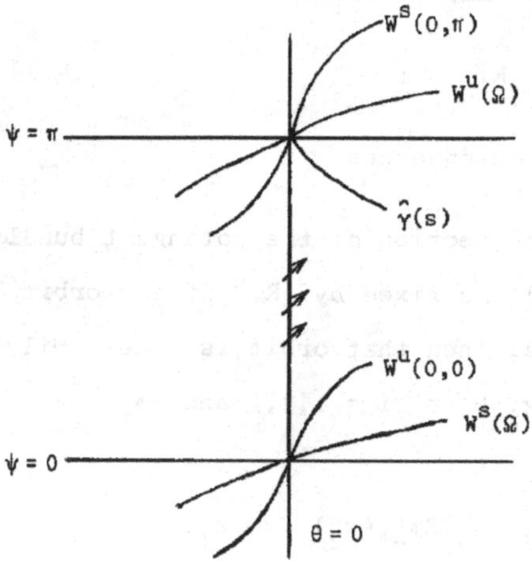

Figure 3. The intersection of the invariant manifolds with T.

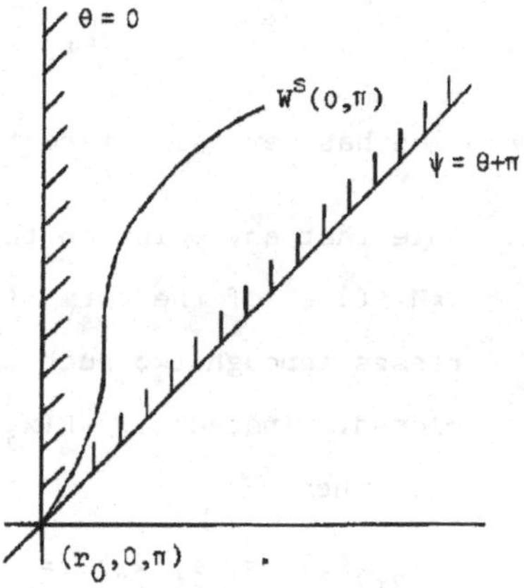

Figure 4: The sector D.

§4. Symmetric Orbits

In this section we use the transversality of the primary bi-collision orbits together with the reversibility of the flow to show that there exist infinitely many long periodic orbits which oscillate about the bi-collision orbits on the q_2-axis. Similar methods will also show that there exist infinitely many long bi-collision orbits. We emphasize that much more can in fact be shown: there is a subsystem of the Anisotropic Kepler problem which is topologically conjugate to a Bernoulli shift on infinitely many symbols. The proof of this, however, is much more difficult and may be found in [4].

A classical mechanical system is called (time-) reversible if the involution $R(q,p) = (q,-p)$ reverses the vector field, that is, $dR(X) = -X$. If the phase space is the cotangent bundle of some manifold M, then R fixes the configuration space variables while reversing the momenta. In terms of the flow φ_t of the vector field X, reversibility implies that each time t map satisfies

$$R\varphi_t = \varphi_{-t}R. \tag{4.1}$$

This relation has several interesting consequences.

1. Note that any point on the zero section of the cotangent bundle T^*M (i.e. of the form $(\underline{q},\underline{0})$) is fixed by R. If any orbit passes through two such points, then that orbit is necessarily closed. Indeed, if $R(x_i) = x_i$ for $i = 1,2$, and $\varphi_t(x_1) = x_2$, then

   $$\varphi_{2t}(x_1) = \varphi_t(x_2) = \varphi_t R x_2 = R\varphi_{-t}(x_2) = x_1.$$

   Such orbits were first noted by Hill in his lunar researches and are now called <u>symmetric periodic orbits</u>.

2. Now suppose that y_1 and y_2 are equilibrium points for a reversible system which satisfy $R(y_1) = y_2$. We observe that R maps $W^s(y_1)$ to $W^u(y_2)$ and $W^u(y_1)$ to $W^s(y_2)$. This follows since, if

   $$\lim_{t\to\infty} \varphi_t(x) = y_1$$

   then

   $$\lim_{t\to\infty} \varphi_{-t}R(x) = R \lim_{t\to\infty} \varphi_t(x) = R(y_1) = y_2.$$

Consequently, if x is fixed by R and also lies in $W^s(y_1)$, then the orbit through x is heteroclinic and lies in $W^s(y_1) \cap W^u(y_2)$. Such orbits are called <u>symmetric heteroclinic orbits</u>.

3. The Anisotropic Kepler problem is reversible in this sense, and the fixed point set of R in each negative energy surface is exactly the oval of zero velocity.

We will use these three observations to prove the existence of infinitely many symmetric periodic and heteroclinic (bi-collision) orbits in the Anisotropic Kepler problem. Note first that the primary bi-collision orbits $\gamma_1^{\pm}$ meet the oval of zero velocity at two points which we denote by q_1^+ and q_1^- respectively. Similarly, $\gamma_2^{\pm}$ meet Z at $q_2^{\pm}$. Hence these orbits are symmetric heteroclinic orbits in the above sense.

Now γ_1^+ lies in the (transverse) intersection of the two dimensional invariant manifolds $W^s(0,\pi)$ and $W^u(0,0)$. We claim that both of these manifolds are also transverse to Z at q_1^+ for $\mu > 1$. To see this we first observe that $R(W^s(0,\pi) = W^u(0,0)$. Hence if $W^s(0,\pi)$ is tangent to Z, then so is $W^u(0,0)$. This, then, would give two independent directions of tangency between $W^s(0,\pi)$ and $W^u(0,0)$: one along γ_1^+ and the other along Z. Since $W^s(0,\pi)$ is transverse to $W^u(0,0)$ at q_1^+, this cannot happen, and hence both $W^s(0,\pi)$ and $W^u(0,0)$ meet Z transversely.

Alternatively, one can use the proof in 3 to show this. Indeed, we showed there that $W^u(0,0)$ met $W^s(\Omega)$ transversely. In the present context, $W^s(\Omega)$ is just the set of orbits which cross Z.

This fact enables us to find infinitely many symmetric closed orbits in each energy level, at least for most $\mu > 9/8$. First consider the sinks in Λ at $(\pm\pi/2, \pm\pi/2)$. From Table I we see that the stable manifolds of both of these sinks are two dimensional and lie entirely in Λ. We may erect small annular transversals $A^{\pm}$ to the flow around each of these sinks. We may further assume that $A^{\pm}$ meet the local stable manifolds to the sinks in circles which we denote by $\sigma^{\pm}$.

Now consider the equilibrium point $(0,\pi)$ in Λ. Its unstable manifold is a curve which lies entirely in Λ. By the results of Section 2, each of the two branches of $W^u(0,\pi)$ is forward asymptotic to a distinct sink for most $\mu > 1$. Hence one branch must die at $(\pi/2,\pi/2)$ and the other at $(-\pi/2,-\pi/2)$. Consequently, one branch of $W^u(0,\pi)$ meets A^+ at a point which we denote by x^+, and the other branch crosses A^- at x^-.

We now follow small pieces of Z under the flow and see how they first intersect $A^{\pm}$. First consider small intervals $B^{\pm}$ in Z centered at $q_2^{\pm}$. Under backward time, these intervals approach the stable manifolds at $(\pm\pi/2, \pm\pi/2)$. For $\mu > 9/8$, the eigenvalues at the sinks are complex, and so $B^{\pm}$ tends to spiral around $W^u(\pm\pi/2, \pm\pi/2)$ as it approaches the stable manifold.

Now if $B^{\pm}$ is chosen small enough, the first points of intersection of backward orbits of points in $B^{\pm}$ with $A^{\pm}$ trace out smooth curves in $A^{\pm}$. In fact, shrinking B^+ further, one may check easily that this curve consists of two smooth spirals, each of which converge (in the C^1 sense) to σ^+. Also, the backward orbits of points in B^- cross A^- along two spirals which converge to σ^-.

We now consider the forward orbit of a small interval C centered at q_1^+ in Z. $C - q_1^+$ consists of two branches $C^{\pm}$. As time changes, one of these branches, say C^+, approaches the branch of $W^u(0,\pi)$ which dies at $(\pi/2,\pi/2)$ and the other accumulates on the branch which dies at $(-\pi/2,-\pi/2)$. Consequently, if C is chosen small enough, then the forward orbits of points in $C^{\pm}$ cut $A^{\pm}$ in a smooth curve which approaches $x^{\pm}$. One may check that in fact this curve is transverse to $\sigma^{\pm}$ in $A^{\pm}$ at $x^{\pm}$. See Figure 5.

Now the existence of symmetric periodic solutions is immediate.

The images of $C^{\pm}$ in $A^{\pm}$ must cross the spiralling images of $B^{\pm}$ at infinitely many distinct points. The orbit of each such point then has two points of intersection with the oval of zero velocity. By our previous remarks, it follows that each such orbit is symmetric periodic.

We remark that the projections of these orbits to the q_1,q_2-plane wind around the q_2-axis many times before hitting the oval of zero velocity near q_1. See Figure 6.

We next turn our attention to the existence of additional symmetric bi-collision orbits in the Anisotropic Kepler problem. These are found in much the same manner as above, hence we only sketch the proof.

First set up annular transversals $D^{\pm}$ to the flow about the sources at $(\pi/2,-\pi/2)$ and $(-\pi/2,\pi/2)$. In forward time, the intervals $B^{\pm}$ above meet $D^{\pm}$ in pairs of spirals, each of which converge smoothly to $D^{\pm} \cap \Lambda$. Now consider $W^s(0,\pi) \cap \Lambda$. This is a curve in Λ consisting of two branches. Again, by the results of Section 2, for most $\mu > 1$, each branch of $W^s(0,\pi) \cap \Lambda$ meets $D^{\pm}$ at two points which we denote by $y^{\pm}$. Now consider all of $W^s(0,\pi)$. This is a two dimensional immersed submanifold. One checks easily that $W^s(0,\pi)$ meets Λ transversely along both branches of $W^s(0,\pi) \cap \Lambda$. As a consequence, $W^s(0,\pi)$ meets $D^{\pm}$ in curves which approach $y^{\pm}$. These curves are transverse to $D^{\pm} \cap \Lambda$ at $y^{\pm}$, and thus must meet the spirals above infinitely often. Each such point of intersection gives rise to an orbit in $W^s(0,\pi)$ which eventually crosses the oval of zero velocity. Such an orbit is therefore a symmetric bi-collision orbit.

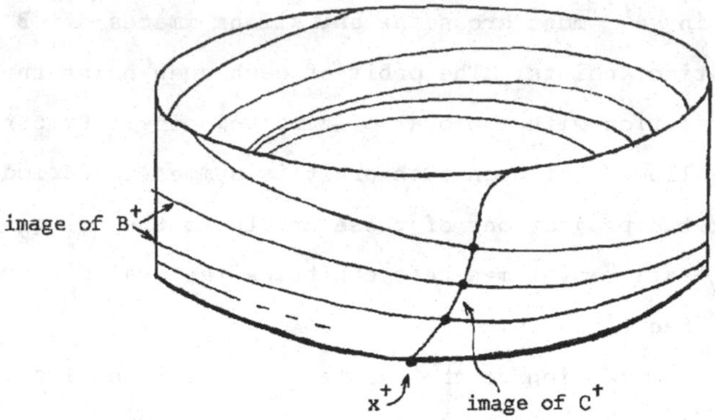

Figure 5. The intersections of B^+ and C^+ in A^+ each give rise to a symmetric periodic orbit.

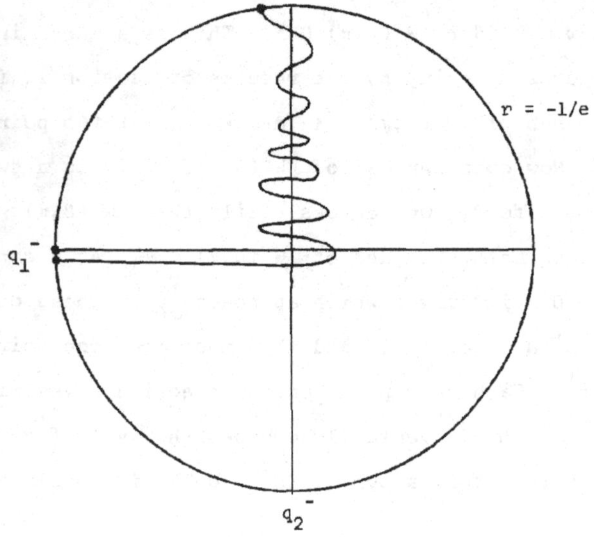

Figure 6. The projection of a typical symmetric closed orbit to configuration space.

REFERENCES

1. C. Conley and R. Easton, Isolated invariant sets and isolating blocks, Trans. AMS 158(1971), 35-61.

2. C. Conley, Some applications of topology in differential equations, Preprint, University of Wisconsin, Madison.

3. R. Devaney, Homoclinic orbits in Hamiltonian systems, J. Diff. Eq. 21(1976), 431-438.

4. R. Devaney, Collision orbits in the Anisotropic Kepler problem, to appear.

5. R. Devaney, Non-regularizability of the Anisotropic Kepler problem, to appear.

6. R. Easton, Regularization of vector fields by surgery, J. Diff. Eq. 10(1971), 92-99.

7. R. Easton, Isolating blocks and symbolic dynamics, J. Diff. Eq. 17 (1975), 96-118.

8. M.C. Gutzwiller, J. Math. Phys. 8(1967), 1979; 10(1969), 1004; 11(1970), 1971; and 12(1971), 343.

9. M.C. Gutzwiller, The anisotropic Kepler problem in two dimensions, J. Math. Phys. 14(1973), 139-152.

10. M.C. Gutzwiller, Bernoulli sequences and trajectories in the anisotropic Kepler problem, to appear.

11. R. McGehee, Triple collision in the collinear three-body problem, Inv. Math. 27(1974), 191-227.

12. R. McGehee, Double collisions for non-Newtonian potentials, to appear.

13. J. Moser, Stable and random motions in dynamical systems, Princeton University Press, Princeton, N.J., 1973.

14. J. Moser, Regularization of Kepler's problem and the averaging method on a manifold, Comm. Pure Appl. Math. 23(1970), 609-636.

15. S. Smale, Differentiable Dynamical Systems, Bull. AMS 73(1967), 747-817.

TUFTS UNIVERSITY

A NOTE ON A DISTALLITY THEOREM
OF C.C. MOORE

by

Douglis Dokken

§0. Introduction

In this paper we show a version of C.C. Moore's Distallity Theorem [4] can be proved without representation theory. Moore shows that if G is a group of affine transformations on an affine space M, G_0 the connected component of the identity in G (G having the compact-open topology) and G/G_0 compact then G is distal on M if and only if $\forall\ g \in G$ the eigenvalues of g are of absolute value one. We prove a similar result for 1-parameter subgroups of affine transformations (infinitesimal affine) acting on a complete connected C^∞-Riemannian manifold (Theorem 1.4). For affine spaces this result implies distallity (Example 1.8).

The results are stated in terms of the linear frame bundle $L(M)$. All connections are assumed to be metric and are named by their connection form ω.

The main theorem is:

> If $\tilde{X}$ is the natural life of X to $L(M)$ and u_p is a frame at p, (p,q) is a distal pair $p \neq q$ in (M,X) if $\omega\tilde{X}(u_p)$ has only pure imaginary eigenvalues and a certain "other condition" is satisfied. This other condition is automatically satisfied for an affine space (Example 1.8).

The preliminaries to the statement of Theorem 1.4 deal with this other condition.

<u>Remark 1.0</u>. Throughout the paper u_p will denote a frame at p

(ordered basis of T_pM written as a row vector) and a, b will denote points in R^n (written as column vectors). A tangent vector, $\sum_i b_i(u_p)_i$ in T_pM will be written as $u_p \cdot b$. In general, the notation of [3] will be used.

Definition 1.1.

(1) Let $f_p: T_pM \to T_pM$ where $f_p(u_p \cdot b) = u_p \cdot \omega\tilde{X}(u_p) \cdot b$.

(2) Let $p_0^{r_0}(x) p_1^{r_1}(x) \ldots p_s^{r_s}(x)$ be the minimal polynomial of f where $p_0(x) = x$ and $p_j(x) = x^2 + a_j^2$, $a_j \neq 0$. Define $E_j' = \mathrm{Ker}(p_j^{r_j}(f_p))$, $E_j = \mathrm{Im}(p_j(f_p)\big|_{E_j'})$ and $E = \oplus_{j=0}^{s} E_j$.

Remark 1.2. (1) For $i = 1,\ldots,s$ there exists a matrix

$$a = \begin{bmatrix} a_1^* & & \\ & \ddots & \\ & & a_s^* \end{bmatrix} \quad \text{where} \quad a_j^* = \begin{bmatrix} \begin{pmatrix} 0 & 1 \\ -a_j^2 & 0 \end{pmatrix} & b & & \bigcirc \\ \ddots & \ddots & \ddots & \\ & \ddots & \ddots & b \\ & & \ddots & c \\ \bigcirc & & & \begin{pmatrix} 0 & a_j \\ -a_j & 0 \end{pmatrix} \end{bmatrix},$$

$$b = \begin{bmatrix} 0 & 0 \\ 1 & 0 \end{bmatrix}, \quad c = \begin{bmatrix} 0 & 0 \\ 1/a_j & -1/a_j \end{bmatrix}.$$ See [2], pp. 95-97.

(2) It is clear from a straightforward calculation that

$$p_j(a_j^*)\begin{bmatrix} \\ v_1 \\ \vdots \\ v_k \\ \\ \end{bmatrix} = \begin{bmatrix} v_3 \\ v_4 \\ \vdots \\ \frac{1}{a_i} v_{k-1} - \frac{1}{a_i} v_k \\ v_k \\ 0 \\ 0 \end{bmatrix}$$

and that

$$\exp(a_j t) = \begin{bmatrix} \bullet\!\!-\!\!-\!\!-\!\!-\!\!-\!\!-\!\!-\!\!- & & \\ & \bullet\!\!-\!\!-\!\!-\!\!-\!\!- & \\ & & \bullet\!\!-\!\!-\!\!- \\ 0 & & \begin{bmatrix} \cos a_j t & \sin a_j t \\ -\sin a_j t & \cos a_j t \end{bmatrix} \end{bmatrix}.$$

Definition 1.3.

(1) Let $B_\varepsilon(E)$ be the ε neighborhood of E.

(2) For fixed p and q in M, define

$$V = \{v \in T_pM \mid \exp_p v = q\}.$$

Theorem 1.4. Let X be an infinitesimal affine transformation on M. and $\omega\tilde{X}(v_p) = a$ have only pure imaginary eigenvalues. If there exists $\varepsilon > 0$ so that the set $V \cap B_\varepsilon(E)$ is finite, then (p,q) is a distal pair.

Remark 1.5. The proof of Theorem 1.4 breaks down into 2 parts:

Part I of the proof handles the case of geodesics associated to $v \in V \setminus V \cap B_{\varepsilon_0}(E)$, $\varepsilon_0 > 0$. The lengths of the images of these geodesics joining p and q do not shrink indefinitely under the action of X. This part of the proof depends upon the fact that the lower

2×2 block in each diagonal block, a_j^*, acts as an isometry on the associated components of v.

Part II handles the case of $v \in V \cap B_\varepsilon(E)$. In the proof of this part we use the Jordan canonical form to measure contracting effects of X on geodesics associated to vectors in $V \cap B_\varepsilon(E)$.

Together these parts allow us to conclude that $\exists\, \varepsilon > 0 \ni \forall\, t \in R$

$$d(X_t(p), X_t(q)) > \varepsilon .$$

<u>Part I of proof</u>: Let X be an infinitesimal affine transformation on M and $\omega\tilde{X}(u_p) = a$ have only pure imaginary eigenvalues. Given $\varepsilon_0 > 0$ there exists $\varepsilon > 0$ such that

$$\forall\, v \text{ in } V \setminus V \cap B_{\varepsilon_0}(E) \text{ and } \forall\, t \in R$$

$$\varepsilon > \int_0^1 \left\| \frac{d}{dt'} \exp_{X_s(p)} t'\, dX_s(p)v \Big|_{t'=t} \right\| dt .$$

<u>Proof</u>: Let $0 < \varepsilon' < \varepsilon_0/2$, then $B_{\varepsilon'}[V \setminus V \cap B_{\varepsilon_0}(E)] \cap B_{\varepsilon'}(E) = \emptyset$. Thus $d(V \setminus V \cap B_{\varepsilon_0}(E), E) > \varepsilon'$. Now $\forall\, v = u_p \cdot \xi_v$ in T_pM,

$$\int_0^1 \left\| \frac{d}{dt'} \exp_{X_s(p)} t'\, dX_s(p)v \Big|_{t'=t} \right\| dt \tag{1.5.1}$$

$$= \int_0^1 \|dX_s(p)v\|\, dt = \|dX_s(p)v\| = \|u_p \cdot e^{\omega\tilde{X}(u_p)s} \cdot \xi_v\| .$$

Choose $u_p \in L(M)$ such that $\omega\tilde{X}(u_p) = a$ is in the form of Remark 1.2 (1).

Let $v \in V \setminus V \cap B_{\varepsilon_0}(E)$. The lower 2×2 block in each diagonal block, a_j^*, of e^{as} acts as an isometry on the associated components of ξ_v. These associated components are (by definition) exactly those which determine $d(V \setminus V \cap B_\varepsilon(E), E)$. Thus, if we choose $w \in E$ such that ξ_w agrees with ξ_v in E and is 0 in remaining components then

$$\|dX_s(p)v\| \geq \|dX_s(p)(v-w)\| \geq d(V\setminus V\cap B_{\varepsilon_0}(E), E) > \varepsilon'.$$

Part II of proof: Assume $V \cap B_\varepsilon(E)$ finite. Let us recall (p,q) is a distal pair if there exist $\varepsilon > 0$ such that $\forall\, t \in R$, $d(X_t(p), X_t(q)) \geq \varepsilon$.

Let $v \in V \cap B_\varepsilon(E)$, $v = u_p \cdot \xi_v$ where u_p is a frame at p and $\xi_v \in R^n$, then

$$\|u_p \cdot e^{at_n} \cdot \xi_v\| \to 0 \text{ as } n \to \infty \text{ if and only if } \|e^{at_n} \cdot \xi_v\| \to 0 \text{ as } n \to \infty.$$

Therefore we show there exists $\varepsilon > 0$ such that $\forall\, t \in R$ $\|e^{at} \cdot \xi_v\| > \varepsilon$ for fixed $v \in V \cap B_\varepsilon(E)$ (see equation 1.5.1).

Let K be a complex $n \times n$ matrix such that $K^{-1}aK$ is in Jordan canonical form.

$(K^{-1}e^{Jt}K)$ is a $\dim(e^{Jt})$-tuple of polynomials times exponentials. Find the upper bound D of the absolute value of the zeros and inflection points of the polynomials in t in the entries of $(K^{-1}e^{at}K)\xi$.

Outside the interval $[-D,D]$ the polynomials are increasing in absolute value so

$$\inf_{t \in R \sim [-D,D]} \|(K^{-1}e^{Jt}K)\xi_v\| \geq \inf_{t \in [-D,D]} \|(K^{-1}e^{Jt}K)\xi_v\| .$$

Since $K^{-1}e^{at}K$ is invertible for every $t \in R$ and $(K^{-1}e^{at}K)\xi_v$ is a continuous function on $[C,D]$ compact, therefore it must assume its minimum value. We have $\inf_{t \in [-D,D]} \|(K^{-1}e^{at}K)\xi_v\| > 0$. By continuity of matrix multiplication there exists $\varepsilon > 0$ such that $\forall\, t \in R$ $\|e^{at}\xi_v\| > \varepsilon$.

Corollary 1.6. Let X and M be as in Theorem 1.4 and V finite, the (p,q) is a distal pair.

Corollary 1.7. Let M be arcwise connected and have a pole p and X be infinitesimal affine on M, with only pure imaginary eigenvalues. If $\pi_1(M)$ is finite, then (M,X) is point distal.

Proof: If (M,P) is a covering space for M, $P^{-1}(q)$ is in 1-1 correspondence with $\pi_1(M,q)/P_*\pi_1(M,q)$, $q \in M$, [6], page 60. By [6], page 47, $\pi_1(M,p_1) = \pi_1(M,p_2) = \pi_1(M)$ for every $p_1,p_2 \in M$ therefore $\pi_1(M,q)/P_*\pi_1(M,q)$ is finite.

Since $\exp_p$ is a covering map by [1], page 170, $V = \exp_p^{-1}(q)$ is finite. Hence by Corollary 1.6, (p,q) is a distal pair $\forall\ q \in M$.

Example 1.8. Let X be infinitesimal affine on R^n then $X(x) = Ax + b$, $A \in gl(n)$ and $b \in R^n$. Using the usual flat Riemannian connection on R^n, $\omega\tilde{X} = A$, where $\tilde{X}$ is the natural lift of X to $L(R^n)$. $\forall\ p,q \in M$, V is a singleton set, therefore if A has only 0 for pure imaginary eigenvalues (M,X) is distal.

Let X be as in Example 1.8, then

$$X_t(x) = e^{At}x + \int_0^t e^{(t-s)A}\, b\, ds.$$

$G = \{X_t(x)\}_{t \in R}$ is a group of affine transformations and if $\omega\tilde{X} = A$ has only 0 and pure imaginary eigenvalues, e^{At} has only eigenvalues of absolute value 1. Therefore, in the special case of 1-parameter affine actions on R^n Moore's result [4] follows from Theorem 1.4.

On the other hand Theorem 1.4 deals with 1-parameter actions on complete connected Riemannian manifolds and is a point distillity result.

BIBLIOGRAPHY

1. N. Hicks, Notes on Differential Geometry, Princeton, N.J., Van Nostrand, 1965.

2. N. Jacobson, Lectures in Abstract Algebra, Vol. II, Van Nostrand, 1953.

3. Kobayashi and Nomizu, Foundations of Differential Geometry, Interscience Tracts in Pure and Applied Mathematics, No. 15, Interscience Publishers, 1963.

4. C.C. Moore, Distal affine transformation groups, Amer. J. Math. 90 (1968), 733-751.

5. W. Perrizo, ω-linear vector fields on manifolds, Trans. Amer. Math. Soc. 196(1974), 289-312.

6. Singer and Thorpe, Lecture Notes on Elementary Topology and Geometry, Scott, Foresman and Co., 1967.

NORTH DAKOTA STATE UNIVERSITY GRADUATE 1975

CHAIN TRANSITIVITY AND THE DOMAIN OF INFLUENCE OF AN INVARIANT SET

by

Robert Easton*

Let (X,d) be a compact metric space and let f be a homeomorphism of X. Iterates of f define an action of the integers on X, a <u>discrete dynamical system</u>. $I \subset X$ is an invariant set (of f) if $f(I) = I$. By the <u>domain of influence of</u> I we mean the set $D(I) = \{x \in X : \alpha(x) \cap I \neq \emptyset$ and $\omega(x) \cap I \neq \emptyset\}$ where $\alpha(x)$ and $\omega(x)$ denote as usual the alpha and omega limit sets of x with respect to the discrete dynamical system generated by f. Orbits in $D(I)$ pass arbitrarily close to I in both the past and the future and thus their behavior might be influenced by the behavior of orbits in I. Notice that $D(I)$ is an invariant set and $D(D(I)) = D(I)$.

The ultimate aim of one studying a dynamical system from a topological point of view is to understand its phase portrait or orbit structure. In most cases this proves too difficult and one tries instead to find and study simple invariant sets such as periodic orbits and to study the orbit structure in neighborhoods of these sets. Many interesting dynamical systems (Hamiltonian systems for example) preserve a given Borel measure μ on X with $\mu(X) = 1$. These systems may be studied from a measure theoretic point of view, and invariant sets of interest to a topologist will often have measure zero. It is my opinion that the study of the orbit structures on such sets may still be important to the measure theoretic study of dynamical systems. It is easy to give examples where I is an invariant set with $\mu(I) = 0$ but $\mu(D(I)) > 0$.

*Partially supported by National Science Foundation (MCS76-84420).

Proposition 1. *Suppose that* μ *is an* f *invariant finite Borel measure on* X *and that* f *is ergodic with respect to* μ. *If* I *is an invariant set, then* $\mu(cl(D(I)) = \mu(X)$.

Proof: Let $\{U_k\}$ be a nested family of neighborhoods of I with $I = \cap_{k=1}^{\infty} U_k$. Then $cl\{x : \omega(x) \cap I = \emptyset\} \subset \cup \{x : f^n(x) \notin U_k \ \forall n \geq 0\}$. Since f is ergodic $\mu(\{x : f^n(x) \notin U_k \ \forall n \geq 0\}) = 0$. Therefore $\mu(cl\{x : \omega(x) \cap I = \emptyset\}) = 0$ and similarly $\mu(cl\{x : \alpha(x) \cap I = \emptyset\}) = 0$. Hence $\mu(cl(D(I)) = \mu(X)$.

Chains or pseudo orbits have been introduced and studied by Rufus Bowen [2] and Charles Conley [3]. An ε-chain is a sequence of points $x_0,\ldots,x_n \in X$ such that $d(f(x_{k-1}),x_k) < \varepsilon$ for $k = 1,\ldots,n$. Define an invariant set I to be *chain transitive* if given $p,q \in I$ and $\varepsilon > 0$ there exists an ε-chain $x_0,\ldots,x_n$ such that $x_0 = p$ and $x_n = q$. Define I to be *strong chain transitive* if given $p,q \in I$ and $\varepsilon > 0$ there exists an ε-chain $x_0,\ldots,x_n$ from p to q such that $\sum_{k=1}^{n} d(f(x_{k-1}),x_k) < \varepsilon$. Such a chain is called a *strong* ε-*chain*.

Proposition 2. *If* I *is strong chain transitive (chain transitive) then so is* $cl(D(I))$.

Proof: Let $p,q \in D(I)$. Let $p_1 \in \omega(p) \cap I$ and $q_1 \in \alpha(q) \cap I$. Choose a strong ε-chain $y_0,\ldots,y_\ell$ from p_1 to q_1. Choose N and $M > 0$ such that $d(f^N(p),p_1) < \varepsilon$ and $d(f^{-M}(q),q_1) < \varepsilon$. Then $p,f(p),\ldots,f^N(p),y_0,\ldots,y_\ell,f^{-M}(q),\ldots,q$ is a strong 3ε-chain from p to q. Thus $D(I)$ is strong chain transitive and this property evidently holds for $cl(D(I))$ also.

Recall that f is ergodic if the only L^1 integrals of f are constant almost everywhere. Define f to be *Lipschitz-ergodic* if any Lipschitz function $G: x \to R^1$ such that $G(f(x)) = G(x) \ \forall x \in X$ is constant.

Proposition 3. _Suppose that_ I _is a closed invariant set on which_ f _is strong chain transitive_. _Then_ f _restricted to_ I _is Lipschitz-ergodic_.

Proof: Let $G: I \to R^1$ be a Lipschitz integral of f with Lipschitz constant k. Let $p,q \in I$ and choose a strong ε-chain $x_0,\ldots,x_\ell$ from p to q. Then $G(q) - G(p) = \sum_{k=1}^{n} G(x_k) - G(x_{k-1})$. Thus

$$\begin{aligned} |G(q) - G(p)| &\leq \sum |G(x_k) - G(x_{k-1})| \\ &\leq \sum |G(x_k) - G(f(x_{k-1}))| \\ &\leq k \sum d(x_k, f(x_{k-1})) \leq k\varepsilon . \end{aligned}$$

Since ε is arbitrary $G(q) = G(p)$ and therefore G is constant on I.

Corollary. _If_ I _is an invariant set on which_ f _is strong chain transitive, then_ f _is Lipschitz ergodic on_ $cl(D(I))$.

The corollary to Proposition 3 illustrates how the behavior on I extends to $D(I)$. It would be interesting to study $D(I)$ where I is a basic set of an axiom A measure preserving diffeomorphism (of a symplectic manifold) whose stable and unstable manifolds intersect transversally along a nondegenerate homoclinic orbit. Is it possible that $cl(D(I))$ always has positive measure? If so this might help explain the existence of ergodic zones observed between invariant tori in numerical experiments [5].

The question of which invariant sets are important in the study of a dynamical system perhaps has no precise answer. Attractors, Morse sets, the nonwandering set and the chain recurrent set are certainly among the important invariant sets. However if the dynamical system preserves a finite Borel measure it is a consequence of Poincaré's recurrence theorem that the chain recurrent set is all of X. In this case it may be justified to consider those (isolated) invariant sets on which the dynamical system is strong chain transitive to be important.

One can associate with an invariant set I a dual invariant set $I^* = \{x : \alpha(x) \cap I = \emptyset$ and $\omega(x) \cap I = \emptyset\}$. I^* consists of those orbits which are bounded away from I. If I is an attractor then I^* is its dual repeller in the sense of Conley [2]. The phase space X is the disjoint union of the sets $D(I)$, I^*, $H(I,I^*)$, $H(I^*,I)$ where $H(I,J) = \{x : \alpha(x) \subset I$ and $\omega(x) \subset J\}$. Notice that $(I^*)^* \subset D(I)$ and that orbits in $D(I) \cap D(I^*)$ oscillate between I and I^*.

Proposition 4. <u>If I is an invariant set and if $cl(I^*) \cap I = \emptyset$ then I^* is an isolated invariant set.</u>

<u>Proof</u>: Choose a neighborhood U of I with $cl(U) \cap I^* = \emptyset$. Then $X - U$ is a compact isolating neighborhood of I^* since if $p \in \partial(X-U) = \partial U$ then the orbit of p is not contained in $X - U$ since if it were, this would imply that $p \in I^*$, which is ruled out by the choice of U.

Notice that if I is a closed invariant set with $I \cap cl(I^*) = \emptyset$ then by this proposition I must be an isolated invariant set. An interesting case which can occur is the case where I, I^*, $H(I,I^*)$ and $H(I^*,I)$ all have measure zero and hence the oscillatory orbits in $D(I) \cap D(I^*)$ have full measure.

So far we have not given a specific example of an invariant set with $\mu(I) = 0$ and $\mu(D(I)) > 0$. Consider the Anosov transformation f of the torus $T = R^2/Z^2$ determined by the integer matrix $\begin{bmatrix} 2 & 1 \\ 1 & 1 \end{bmatrix}$. f preserves Lebesgue measure μ on T. Let $I = \pi(0)$ where $\pi\colon R^2 \to R^2/Z^2$ is the quotient projection. I is a hyperbolic fixed point of f whose stable and unstable manifolds intersect in a dense subset of T. From its definition $D(I)$ contains this set and therefore it follows that f is strong chain transitive on all of T. Of course it is known that f is ergodic and in fact f is measure theoretically isomorphic to a Bernoulli shift [1]. However these results are difficult to establish whereas the strong chain transitivity and hence Lipschitz ergodicity of f is quite simple to verify.

I want to propose further examples which seem to have some similarity to Anosov transformations. Let $L_1 = \{(x,y) \in R^2 : 0 \leq x \leq 1/2\}$ and $L_2 = \{(x,y) \in R^2 : 0 \leq y \leq 1/2\}$ and let $A_k = \pi(L_k)$ where $\pi: R^2 \to R^2/Z^2$ is the quotient projection. Let $X = A_1 \cup A_2$. X consists of two overlapping annuli in the torus R^2/Z^2. Choose $\beta,\gamma: [0,1/2] \to [0,\infty)$ smooth functions with positive derivatives on $(0,1/2)$, with $\beta(0) = \gamma(0) = 0$ and with $\beta(1/2)$ and $\gamma(1/2)$ integers. Define twist maps on the annuli A_1 and A_2 by $f_1(x,y) = (x+\beta(y),y)$ and $f_2(x,y) = (x,y+\gamma(x))$. Extend these maps to be the identity off A_1 and A_2 respectively. Let f be the composition of twist mappings $f = f_2 \circ f_1$. Each of the maps f_1 and f_2 preserves Lebesgue measure and the dynamical systems generated by f_1 and by f_2 are easy to understand. f_1 preserves the curves y = constant and an orbit is dense in such a curve if and only if $\beta(y)$ is irrational. However I conjecture that the orbit structure of f is quite complicated and that f is at least Lipschitz ergodic.

If $(x,y) \in f_1^{-1}(A_1 \cap A_2)$ then

$$f(x,y) = (x+\beta(y),\ y+\gamma(x+\beta(y))).$$

Let $v = \text{int}\, f^{-1}(A_1 \cap A_2)$. For $(x,y) \in V$,

$$df(x,y) = \begin{pmatrix} 1 & \beta'(y) \\ \gamma'(x+\beta(y)) & 1+\gamma'(x+\beta(y))\beta'(y) \end{pmatrix}.$$

$df(x,y)$ has eigenvalues $1 + 1/2\{\beta'\gamma' \pm [(\beta'\gamma')^2 + 4\beta'\gamma]^{1/2}\}$. Since the derivatives β' and γ' are positive it follows that $df(x,y)$ is hyperbolic. Devaney [4] has shown that V contains a basic set I

on which f is topologically conjugate to a subshift of finite type.

Let

$$Q_1 = \{(x,y) \in A_1 : \beta(y) \text{ is rational}\}$$

$$Q_2 = \{(x,y) \in A_2 : \gamma(y) \text{ is rational}\}$$

$$K = X - \bigcup_{-\infty}^{\infty} f^n(Q_1 \cup Q_2).$$

Then $\mu(K) = \mu(X)$.

Proposition 5. <u>If</u> $(x_0,y_0) \in K$ <u>then there exists</u> $n > 0$ <u>such that</u> $f^n(x_0,y_0) \in V$. <u>Thus the orbits of points in</u> K <u>intersect</u> V <u>infinitely often</u>.

<u>Proof</u>: <u>Case 1</u>: Let $(x_0,y_0) \in K$ and suppose that $(x_1,y_1) = f(x_0,y_0) \in A_1$ and that $f^k(x_0,y_0) \notin V$ for each $k > 0$. Then $f^2(x_0,y_0) = f_1(x_1,y_1) = (x_1+\beta(y_1),y_1)$ and in general $f^{k+1}(x_0,y_0) = f_1{}^k(x_1,y_1) = (x_1+k\beta(y_1),y_1)$. This implies that the positive orbit of (x_0,y_0) is dense in the circle $\{y = y_1\}$. But this circle intersects V which gives a contradiction.

<u>Case 2</u>: Let $(x_0,y_0) \in K$ and suppose that $(x_1,y_1) = f(x_0,y_0) \in A_2$ and that $f^n(x_0,y_0) \notin V$ for each $k > 0$. Then $f^2(x_0,y_0) = f_2(x_1,y_1) = (x_1,y_1+\gamma(x_1))$, $f^2(x_0,y_0) \notin A_1$ by Case 1. Therefore $f^3(x_0,y_0) = f^2(x_1,y_1) = (x_1y_1+2\gamma(x_1))$ and in general $f^{k+1}(x_0,y_0) = (x_1,y_1+k\gamma(x_1))$. This implies that the positive orbit of (x_0,y_0) is dense in the circle $x = x_1$. But this circle must intersect V which gives a contradiction.

Let $p = \pi(0,0)$ and $q = \pi(0,1/2)$. Define $W^u(p) = \{w \in X : f^n(w) \to p \text{ as } n \to -\infty\}$, $W^s(q) = \{w \in X : f^n(w) \to q \text{ as } n \to +\infty\}$. I conjecture that $W^u(p) \cap W^s(p)$ is dense in X. Since there exist strong ε-chains from q to p for each ε it would follow that f is strong chain transitive and hence Lipschitz ergodic on X.

I want to conclude by describing some variations of the examples discussed above. First one might consider a double twist of the torus T defined by $f(x,y) = (x+\beta(y),\ y+\gamma(x+\beta(y)))$ where β and γ are increasing functions on $[0,1]$ with $\beta(0)$, $\beta(1)$, $\gamma(0)$, $\gamma(1)$ integers. In some cases this gives an Anosov transformation. A discrete analogue of the double twist is to consider a rectangular array of points $R = \{(x,y) \in Z \times Z : 1 \le x \le m,\ 1 \le y \le n\}$. Define a permutation f of R by $f(x,y) = (x_1,y_1)$ where $x_1 = x + \beta(y) \bmod m$ and $y_2 = y + \gamma(x+\beta(y)) \bmod n$ with β and γ integer valued increasing functions on $[1,\ldots,m]$ and $[1,\ldots,n]$ respectively. Ergodicity of f in this case means that f is a cyclic permutation of R. It is an interesting combinatorial problem to determine conditions on n, m, β, γ which imply that f is ergodic.

Finally the composition of twist maps can occur as Poincaré maps of flows. Consider the space X consisting of two overlapping annuli in R^2/Z^2 as before. Let $Y = X \times S^1$ and consider the flow φ on Y defined by $\varphi(x,y,\theta) = (x,y,\theta+t)$. Modify this flow as follows. Let

$$R_1 = X \times \{1/3\pi \le \theta \le 2/3\pi\}$$

$$R_2 = X \times \{4/3\pi \le \theta \le 5/3\pi\}$$

and replace the vector field $\dot\theta = 1$ which generates φ inside R_1 and R_2 by the smooth vector fields V_k given by $\dot x = u_k(y,\theta)$, $\dot y = v_k(x,\theta)$, and $\dot\theta = w_k(x,y,\theta)$ for $k = 1,2$. Choose smooth functions u_k, v_k, w_k with the following properties:

(1) u_k, v_k vanish on ∂R_k and w_k is equal to 1 on ∂R_k.

(2) $v_1 \equiv 0$, $u_2 \equiv 0$, $u_1 \ge 0$, $v_2 \ge 0$.

(3) $u_1(1/4,\pi/2) = 1 = v_2(1/4,3\pi/2)$.

(4) $w_1 \ge 0$ and $w_1 = 0$ on the set $y = 1/4$, $\theta = \pi/2$.

(5) $w_2 \ge 0$ and $w_2 = 0$ on the set $x = 1/4$, $\theta = 3\pi/2$.

Let ψ be the flow generated by the modified vector field on X. $X \times 0$ is a surface of section for this flow and the Poincaré map $f: X \times \{0\} \to X \times \{0\}$ determined by the flow has the form $(x,y) \to (x+\beta(y),\ y+\gamma(x+\beta(y)))$ where β and γ are smooth functions defined on $[0,1/4) \cup (1/4,1/2]$ with graphs of the type pictured in Figure 1. Here the twist maps have an infinite twist and f is undefined on the circles $y = 1/4$, $\theta = 0$ and $x = 1/4$, $\theta = 0$.

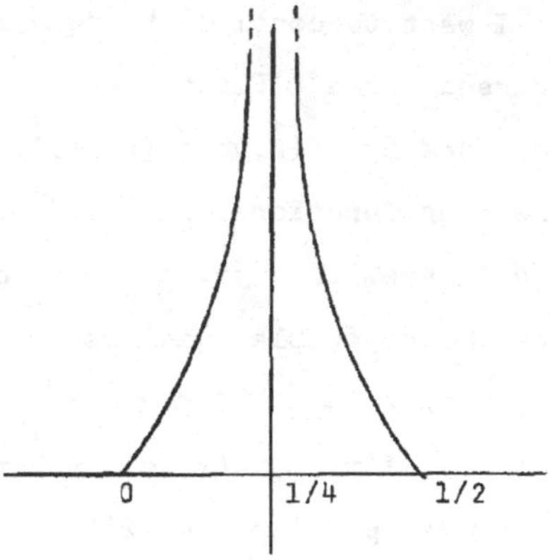

Figure 1

REFERENCES

1. R.L. Adler and B. Weiss, Similarity of automorphisms of the torus, Memoirs of the A.M.S., Number 98.

2. R. Bowen, On Axiom A diffeomorphisms, C.B.M.S. Regional Conference Series in Mathematics.

3. C. Conley, Isolated invariant sets and the Morse Index, C.B.M.S. Regional Conference Series in Mathematics.

4. R. Devaney, Subshifts of finite type in linked twist mappings.

5. M. Henon and C. Heiles, The applicability of the third integral of motion; some numerical experiments, The Astronomical Journal 69 (1964), 73-79.

UNIVERSITY OF COLORADO

COHOMOLOGY OF FLOWS

by

Robert Ellis*

§0. Introduction

In this paper I would like to lay the groundwork for the application of the general theory of the cohomology of groups to problems in topological dynamics. Work along these lines was begun by Peterson in [4] who used the fact that what is generally called a cocycle on a flow (X,T) to a compact abelian group K is just a one-cocycle of the group T with coefficients in the T-module, $C(X,K)$ of continuous K-valued functions on X. In [1] the study of cocycles on minimal sets was linked up with the algebraic theory expounded in [2]. The reason the latter is applicable is that to each cocycle σ on (X,T) to K there is associated a continuous function f_σ from βT to K such that f_σ is homomorphism from the group of X to K. The cocycle σ may also be viewed as a one-cocycle of T with coefficients in $C(\beta T,K)$, and as such it is just the coboundary of f_σ; i.e. $\sigma = df_\sigma$. However it was not clear why f_σ was a homomorphism when restricted to the group of X nor what was to be expected in higher dimensions. The solution to this problem is obtained by considering the cohomology of T with coefficients in K alongside the cohomology of T with coefficients in $C(X,K)$. It turns out that f_σ is a cocycle in the former complex when restricted to the group of X. (This statement is made precise in Proposition 2.9.)

It also became clear in [1] that for most applications the restriction that K be abelian is too stringent. Consequently I have shown how one might define a cohomology theory for non-abelian K in the context of minimal sets. Of course, the two theories coincide

*Partially supported by National Science Foundation (MPS75-05250).

when K is abelian.

Section 1 is devoted to the development of the necessary formalism, Section 2 to the definition of the cohomology theory, and Section 3 to the derivation of some exact sequences. In a future paper I hope to use these results in order to classify the A-supplements of F in B where $A \subset F \subset B$ (see [1] for definitions and details).

0.1. Standing notation

In general I shall use the notation employed in [2] without further comment.

§1. The Formalism

In this section two cochain complexes pertinent to topological dynamics are introduced, and some basic formulas derived.

Definition 1.1. Let (T,L) be a transformation group where the phase space L is a discrete group and the map $x \to tx : L \to L$ is a homomorphism $(t \in T)$. Then following [3] I denote by $C^n(L)$ or C^n the set of functions $c: T^n \to L$ such that $c(t_1,\ldots,t_n) = 0$ if one of the t's equals the identity e of T.

Let $c: T^n \to L$. Then dc will denote the map from T^{n+1} to L such that

$$
\begin{aligned}
dc(t_1,\ldots,t_{n+1}) \;=\; (-1)^{n+1}[t_1c(t_2,\ldots,t_{n+1}) \\
+ \sum_{i=1}^{n} (-1)^i c(t_1,\ldots,t_\ell t_{\ell+1},\ldots,t_{n+1}) \\
+ (-1)^{n+1}c(t_1,\ldots,t_n)].
\end{aligned}
$$

Remarks 1.2. 1. Even though the group operation in L is written additively I do not assume that L is abelian.

2. It is easy to verify that d maps C^n into C^{n+1}. However since L need not be abelian, d is in general not a homomorphism nor is $d^2 = 0$.

3. In this paper I shall be interested in two particular values for L. The first of these is $L = K^T$. The action of T on K^T is given by $(tf)(s) = f(st)$ $(f \in K^T,\ s,t \in T)$.

Let $c \in C^n(K^T)$. Then I shall write $c(s,t_1,\ldots,t_n)$ instead of $c(t_1,\ldots,t_n)(s)$. Thus $(tc)(t_1,\ldots,t_n)(s)$ becomes $c(st,t_1,\ldots,t_n)$. This complex which I denote (C_1^n,d_1) or simply (C_1^n) may be described: C_1^n is the set of functions $c\colon T^{n+1} \to K$ such that $c(r,t_1,\ldots,t_n) = 0$ when one of the t's equals e, and $d_1\colon C_1^n \to C_1^{n+1}$ is such that

$$d_1c(t_1,\ldots,t_{n+2}) = (-1)^{n+1}\Big[\sum_{\ell=1}^{n+1} (-1)^{\ell-1}c(t_1,\ldots,t_\ell t_{\ell+1},\ldots,t_{n+2}) + (-1)^{n+1}c(t_1,\ldots,t_{n+1})\Big].$$

The complex (C_1^n,d_1) with K abelian is essentially the one used by Peterson in [4].

4. The other complex which I denote (C_2^n,d_2) is obtained by setting $L = K$ and letting T act trivially. Thus C_2^n is the set of functions $c\colon T^n \to K$ such that $c(t_1,\ldots,t_n) = 0$ if one of the t's is the identity of T and $d_2\colon C_2^n \to C_2^{n+1}$ is such that

$$d_2c(t_1,\ldots,t_{n+1}) = (-1)^{n+1}\Big[c(t_2,\ldots,t_{n+1}) + \sum_{\ell=1}^{n} (-1)^{i}c(t_1,\ldots,t_\ell t_{\ell+1},\ldots,t_{n+1}) + (-1)^{n+1}c(t_1,\ldots,t_n)\Big].$$

5. The above formulas will be used to define d_1c and d_2c for arbitrary functions $c\colon T^n \to K$.

<u>Definition 1.3</u>. The operators δ and ε. Let $c\colon T^n \to K$. Then δc is the map $(t_1,\ldots,t_{n-1}) \to c(e,t_1,\ldots,t_{n-1}) : T^{n-1} \to K$ and εc the map $(t_1,\ldots,t_{n+1}) \to c(t_2,\ldots,t_{n+1}) : T^{n+1} \to K$.

Lemma 1.4. Let c, c_1, c_2 be maps of T^n to K. Then:

1. $\delta(c_1+c_2) = \delta c_1 + \delta c_2$; 2. $\delta(-c) = -\delta c$; 3. $\varepsilon(c_1+c_2) = \varepsilon c_1 + \varepsilon c_2$;

4. $\varepsilon(-c) = -\varepsilon c$; 5. $\delta\varepsilon c = c$; 6. $\delta^2 c = 0$.

Proof. Immediate.

The proof of the following proposition involves a straightforward but tedious computation and so will be omitted.

Proposition 1.5. Let $c: T^{n+1} \to K$. Then

1. $(-1)^n d_2 c = \varepsilon c + (-1)^{n+1} d_1(-c)$,

2. $(-1)^{n+1} \delta d_1 c + (-1)^n d_2 \delta c = c - \varepsilon\delta c$,

3. $(-1)^{n+1} \delta d_1 c + (-1)^{n+1} d_1 \delta(-c) = c$.

(Again the reader is cautioned that $d_i(-c)$ need not equal $-d_i(c)$, $i = 1,2$.)

Lemma 1.6. The operator d_1 is injective when restricted to $C_2^n \subset C_1^{n-1}$.

Proof. Let $c_1, c_2 \in C_2^n$ with $d_1 c_1 = d_1 c_2$. Since $\delta(-c_1) = -\delta c_1 = 0 = -\delta c_2 = \delta(-c_2)$, 3 of 1.5 gives $c_1 = (-1)^n \delta d_1 c_1 = (-1)^n \delta d_1 c_2 = c_2$.

Definition 1.7. The set of n-cocycles on T with coefficients in K (denoted $Z^n(T,K)$ or simply $Z^n(K)$) is the image of C_2^n in C_1^n under the map d_1.

Let $z_1, z_2 \in Z^n(T,K)$. Then by 1.6 there exist unique elements c_1, c_2 in C_2^n with $d_1 c_i = z_i$, $i = 1,2$. Since $c_1 + c_2 \in C_2^n$, $d_1(c_1+c_2) \in Z^n(T,K)$. We define $z_1 \oplus z_2$ to be $d_1(c_1+c_2)$.

Proposition 1.8. $(Z^n(K)\oplus)$ is a group with identity 0 and where the inverse of $z = d_1 c$ (denoted $\ominus z$) is given by $d_1(-c)$.

<u>Proof</u>. Let $z_i = d_1c_i \in Z^n(K)$, $i = 1,2,3$. Then $(z_1 \oplus z_2) \oplus z_3 = d_1((c_1+c_2)+c_3) = d_1(c_1+(c_2+c_3)) = z_1 \oplus (z_2 \oplus z_3)$. Thus $\oplus$ is associative.

Since $d_1 0 = 0$, $z \oplus 0 = z = 0 \oplus z$ $(z \in Z^n(K))$.

Finally let $z = d_1c \in Z^n(K)$. Then $z \oplus d_1(-c) = d_1(c+(-c)) = d_1(0) = 0$. The proof is completed.

Proposition 1.9. <u>Let</u> $\bar{\delta}c = (-1)^n \delta c$ <u>for all maps</u> c <u>of</u> T^{n+1} <u>to</u> K. <u>Then</u>:

1. $\bar{\delta}(z_1 \oplus z_2) = \bar{\delta}(z_1) + \bar{\delta}(z_2)$,

2. $\bar{\delta}(\theta z) = -\delta(z)$,

<u>for all</u> z_1, z_2, z <u>in</u> $Z^n(K)$, <u>and</u>

3. $\bar{\delta}$ <u>is injective on</u> $Z^n(K)$.

<u>Proof</u>. 1. Let $z_i = d_1c_i$, with $c_i \in C_2^n$, $i = 1,2$. Then applying 3 of 1.5 to c_1, c_2 and $c_1 + c_2$ gives $\bar{\delta}d_1c_i = c_i$ and $\bar{\delta}d_1(c_1+c_2) = c_1 + c_2$, $i = 1,2$. (Recall that $\delta c = 0$, $(c \in C_2^n)$.) In other words $\bar{\delta}z_i = c_i$ and $\bar{\delta}(z_1 \oplus z_2) = c_1 + c_2$ $(i = 1,2)$, whence $\bar{\delta}z_1 + \bar{\delta}z_2 = \bar{\delta}(z_1 \oplus z_2)$.

2. Follows from 1 and the fact that $\bar{\delta}0 = 0$.

3. Let $z_i = d_1c_i \in Z^n(K)$ with $\bar{\delta}z_1 = \bar{\delta}z_2$ $(i = 1,2)$. Then as in 1 above, $c_1 = \bar{\delta}z_1 = \bar{\delta}z_2 = c_2$ whence $z_1 = d_1c_1 = d_1c_2 = z_2$.

Corollary 1.10. <u>Let</u> z_1, z_2 <u>and</u> $z_1 + z_2$ <u>be in</u> $Z^n(K)$. <u>Then</u> $z_1 \oplus z_2 = z_1 + z_2$.

<u>Proof</u>. By 1 of 1.4 $\delta(z_1+z_2) = \delta(z_1) + \delta(z_2)$. Hence $\bar{\delta}(z_1+z_2) = \bar{\delta}(z_1) + \bar{\delta}(z_2) = \bar{\delta}(z_1 \oplus z_2)$ whence $z_1 + z_2 = z_1 \oplus z_2$ by 3 of 1.9.

Proposition 1.11. <u>Let</u> K <u>be abelian</u>. <u>Then</u> $Z^n(K) = \{c \mid c \in C_1^n, d_1c = 0\}$ <u>and</u> $z_1 + z_2 = z_1 \oplus z_2$ $(z_1, z_2 \in Z^n(K))$. Thus in the abelian case definition 1.7 coincides with the usual one.

Proof. Let $z = d_1 c \in Z^n(K)$. Then $d_1 z = d_1^2 c = 0$ since K is abelian.

Conversely let $c \in C_1^n$ with $d_1 c = 0$. Then 3 of 1.5 implies that $c = d_1((-1)^n \delta c)$. Since $\delta c \in C_2^n$, $c \in Z^n(K)$.

Finally, $z_1 + z_2 = z_1 \oplus z_2$ $(z_1, z_2 \in Z^n(K))$ by 1.10.

§2. The Cohomology of a Flow

In this section I define the n-cocycles, n-coboundaries and n-cohomology of a T-subalgebra A of $\mathfrak{C}(u)$.

Lemma 2.1. Let $c: T^n \to K$. Then there exists a unique map $\bar{c}: (\beta T)^n \to K$ such that $\bar{c}|T^n = c$ and the maps $y \to \bar{c}(x_1,\ldots,x_r,y,s_{r+2},\ldots,s_n) : \beta T \to K$ are continuous for all $x_1,\ldots,x_r \in \beta T$, $s_{r+2},\ldots,s_n \in T$ and $1 \le r \le n-1$.

Proof. It is clear that if such a map $\bar{c}$ exists, then it is unique.

Let $x \in \beta T$ and $s_2,\ldots,s_n \in T$. Set $c_1(x,s_2,\ldots,s_n)$ equal to the value at x of the continuous extension to βT of the map $t \to c(t,s_2,\ldots,s_n) : T \to K$. Then $c_1: \beta T \times T^{n-1} \to K$ with $c_1|T^n = c$ and $x \to c_1(x,s_2,\ldots,s_n) : \beta T \to K$ continuous for all $s_2,\ldots,s_n \in \beta T$.

Now assume c_r defined for $1 \le r \le k < n$ such that $c_r: (\beta T)^r \times T^{n-r} \to K$ with $c_r|T^n = c$ and $y \to c_r(x_1,\ldots,x_q,y,s_{q+2},\ldots,s_n) : \beta T \to K$ continuous for all $x_1,\ldots,x_q \in \beta T$ and $s_{q+2},\ldots,s_n \in T$, where $1 \le q \le r-1$.

Let $x_1,\ldots,x_k,x_{k+1} \in \beta T$; $s_{k+2},\ldots,s_n \in T$, and set $c_{k+1}(x_1,\ldots,x_k,x_{k+1},s_{k+2},\ldots,s_n)$ equal to the value at x_{k+1} of the continuous extension to βT of the map $t \to c_k(x_k,\ldots,x_k,t,s_{k+2},\ldots,s_n)$: $T \to K$.

The map $\bar{c}$ is just c_n.

Remarks 2.2. 1. If $c: T^n \to K$, then the map $\bar{c}$ will henceforth also be denoted c.

2. The continuity condition of 2.1 implies that the formulas derived in Section 1 are valid for the extended functions.

Definition 2.3. Let $c \in C_1^n(K)$. Then we write $x \equiv y(c)$ if $c(xz,t_1,\ldots,t_n) = c(yz,t_1,\ldots,t_n)$ $(z \in \beta T,\ t_1,\ldots,t_n \in T)$.

Remarks 2.4. 1. It follows immediately from 2.1 that $x \equiv y(c)$ if and only if $c(xz,z_1,\ldots,z_n) = c(yz,z_1,\ldots,z_n)$ $(z,z_1,\ldots,z_n \in \beta T)$.

2. For $c \in C_1^n$, 2.1 defines a closed invariant equivalence relation on βT. The space of c (denoted $sp(c)$) is defined to be the image of M in $\beta T/(c)$, and $al(c)$ the corresponding T-subalgebra of $\mathfrak{A}(u)$.

3. That $al(c) \subset A$ for some T-subalgebra A of $\mathfrak{A}(u)$ is equivalent to the fact that c induces a continuous map

$$(x|A,t) \to c(x,t) : |A| \times T^n \to K$$

where $|A|$ is the Gelfand space of the algebra A.

Lemma 2.5. Let $c \in C_1^n$ and $z \in Z^n(K)$. Then

(i) $al(dc) \subset al(c)$, and

(ii) $al(z) \subset al(\delta z)$.

Proof. (i) follows directly from the definitions.

(ii) Let $z = d_1 r$ with $r \in C_2^n$. Since $\delta r = 0$, 3 of 1.5 implies that $(-1)^n \delta z = r$. Thus $al(\delta z) = al(r) \supset al(z)$ by (1).

Definition 2.6. Let A be a T-subalgebra of $\mathfrak{A}(u)$. Then the n-cocycles of A with coefficients in K is the set,

$$Z^n(A;K) = \{z \mid z \in Z^n(K),\ al(z) \subset A\},$$

and the n-coboundaries of A with coefficients in K is the set

$$B^n(A;K) = \{z \mid z \in Z^n(K) \text{ and } al(\delta z) \subset A\}.$$

Note that $B^n(A;K) \subset Z^n(A;K)$ by (ii) of 2.5.

When there is no chance of confusion I shall write $Z^n(A)$ and $B^n(A)$ instead of $Z^n(A;K)$ and $B^n(A;K)$.

Lemma 2.7. $B^n(A)$ _is a subgroup of_ $Z^n(K)$.

Proof. Let $z,z_1,z_2 \in B^n(A)$ and $\bar{\delta} = (-1)^n\delta$. Then $al(\delta(z_1 \oplus z_2)) = al(\bar{\delta}(z_1 \oplus z_2)) = al(\bar{\delta}(z_1) + \bar{\delta}(z_2)) \subset A$ (1 of 1.9).

Also $\bar{\delta}(\theta z) = -\bar{\delta}(z)$ implies that $al(\delta(\theta z)) = al(\bar{\delta}(\theta z)) = al(\bar{\delta}z) \subset A$.

Remarks 2.8. 1. In general $Z^n(A)$ is not a subgroup of $Z(K)$. (See 2 of 2.10.) Thus in order to define the nth order cohomology $H^n(A;K)$ of A we must use the subgroup $\bar{Z}^n(A)$ of $Z^n(K)$ generated by $Z^n(A)$ and set $H^n(A) = H^n(A;K) = \bar{Z}^n(A:K)/B^n(A:K) = \{z \oplus B^n(A;K) \mid z \in \bar{Z}^n(A;K)\}$.

We could of course have defined an equivalence relation R on $Z^n(A)$ by setting $(z_1,z_2) \in R$ if $z_1 = z_2 \oplus b$ for some $b \in B^n(A)$ $(z_1,z_2 \in Z^n(A))$ and then defined $H^n(A;K)$ to be $Z^n(A)/R$. This is what was done in [1], but the definition using $\bar{Z}^n(A)$ is needed for the consideration of exact sequences. (See Section 3.)

2. When K is abelian $z_1 \oplus z_2 = z_1 + z_2$ and so $al(z_1 \oplus z_2) \subset A$. Thus in this case $Z^n(A)$ is a group and the two definitions of $H^n(A)$ given above coincide with the usual one [4].

Proposition 2.9. _Let_ $z \in Z^n(K)$ _and set_ $\hat{\delta}z = (-1)^{n+1}\delta z$. _Then_

(1) $(d_2\hat{\delta}z)(x_1,\dots,x_{n+1})$

$$= \begin{cases} z(e,x_2,\dots,x_{n+1}) - z(x_1,\dots,x_{n+1}) & \text{for } n \text{ odd} \\ -z(x_1,\dots,x_{n+1}) + z(e,x_2,\dots,x_{n+1}) & \text{for } n \text{ even.} \end{cases}$$

(2) $z \in Z^n(A)$ _if and only if_ $al(\hat{\delta}z) \subset A$.

(3) $z \in Z^n(A)$ _implies_ $(d_2\hat{\delta}z)(a,t_1,\dots,t_n) = 0$ $(a \in \mathfrak{g}(A)$ _and_

$t_1,\ldots,t_n \in T$).

Proof. 1. Let $z = d_1c$ for some $c \in C_2^n$. Since $0 = \delta c = -\delta(-c)$, 3 of 1.5 shows that $(-1)^n\delta z = c$.

Suppose n is odd. Then $\hat{\delta}z = \delta z = -c$ and $\delta d_1 z + d_1\delta(-z) = z$ by 1.5. Since $\delta(-z) = c$ and $d_1c = z$, $\delta d_1 z = 0$. Now 2 of 1.5 gives: (i) $-d_2\delta z = z - \varepsilon\delta z$.

Now suppose n is even. Then $\hat{\delta}z = -\delta z = -c$ and $-\delta d_1(-z) - d_1\delta z = -z$ by 1.5. Since $\delta z = c$ and $d_1c = z$, $\delta d_1(-z) = 0$. Again using 2 of 1.5 we get: (ii) $d_2\delta(-z) = -z + \varepsilon\delta z$.

Formulas (i) and (ii) give (1).

Statements (2) and (3) follow immediately from (1). The proof is completed.

Remarks 2.10. 1. Let $z \in Z^1(A)$. Then $\hat{\delta}z(x) = (-1)^2\delta z(x) = z(e,x)$. Thus $\hat{\delta}z$ is just what was called the function associated with the cocycle z in [1].

Statement (3) of 2.9 says that $\hat{\delta}z$ restricted to the group A, of A is a d_2-cocycle. This is just the condition that $\hat{\delta}z$ be a homomorphism when restricted to A [1]. Thus (3) is a generalization to higher dimensions of the basic fact exploited in [1], namely that the function associated with a cocycle is a homomorphism when restricted to the group of the cocycle.

2. With the same notation as in 1 let us assume that θz is also in $Z'(A;K)$. Then $\hat{\delta}(\theta z) = -\hat{\delta}z$ is also a homomorphism of A into K. Thus $-(\hat{\delta}z(a)+\hat{\delta}z(b)) = -(\hat{\delta}z(ab)) = (-\hat{\delta}z)(ab) = -\hat{\delta}z(a) - \hat{\delta}z(b)$; from which it follows that $\hat{\delta}z(a) + \hat{\delta}z(b) = \hat{\delta}z(b) + \hat{\delta}z(a)$ $(a,b \in A)$. Thus z and $\theta z \in Z^n(A)$ implies that $\operatorname{im}\hat{\delta}z$ is an abelian subgroup of K.

Now let $|A|$ be 0-dimensional and B a τ-closed normal subgroup of A such that $A^{\#} \subset B \subset A$ with A/B not abelian. Then there exists $z \in Z'(A;A/B)$ with $\hat{\delta}z\colon A \to A/B$ the projection [1]. Hence

$\theta z \notin Z^1(A;A/B)$.

§3. Some Exact Sequences

In this section the two types of exact sequences that occur naturally in the cohomology of groups are discussed. The first is the cohomology sequence of a pair, the main problem being to extend the results of the abelian to the non-abelian case. The other sequence results from varying the coefficient group. Here even the abelian case presents problems because the functions involved need not be continuous.

As with much of the work on exact sequences the proofs are direct but tedious. I shall therefore content myself with giving the definitions, stating the theorems, and pointing out the special problems that occur in the present situation.

Lemma 3.1. Let $z_1 + r_1 = z_2 + r_2$ where $z_i \in Z^n(K)$ and $r_i \in C_2^{n+1}(K)$ $(i = 1,2)$. Then $z_1 = z_2$ and $r_1 = r_2$.

Proof. Let $z_i = d_1 c_i$ with $c_i \in C_2^n$ $(i = 1,2)$. Then by 1.5 $(-1)^n \delta z_i = c_i$ $(i = 1,2)$. Hence $(-1)^n c_1 = \delta z_1 = \delta(z_1+r_1) = \delta(z_2+r_2) = \delta z_2 = (-1)^n c_2$. Thus, $c_1 = c_2$, $z_1 = d_1 c_1 = d_1 c_2 = z_2$ and $r_1 = -z_1 + (z_1+r_1) = -z_2 + (z_2+r_2) = r_2$.

Lemma 3.1 allows us to provide $Z^n(K) + C_2^{n+1}$ with a group operation $\oplus$, namely $c_1 \oplus c_2 = z_1 \oplus z_2 + r_1 + r_2$ for $c_i = z_i + r_i \in Z^n(K) + C_2^{n+1}$ $(i = 1,2)$. With this operation $Z^n(K) + C_2^{n+1}$ becomes the direct sum of the subgroups $Z^n(K)$ and C_2^{n+1}.

Now let F be a T-subalgebra of A and set $C_2^{n+1}(A,F) = \{r \mid r \in C_2^{n+1},\ al(r) \subset A,\ al(dr) \subset F\}$. Then $C_2^{n+1}(A,F)$ is a subgroup of C_2^{n+1}.

Definition 3.2. Let F be a T-subalgebra of A. Then an n-cocycle of A modulo F is an element of the set $Z^n(A,F) = Z^n(A) + C_2^{n+1}(A,F)$.

As before $Z^n(A,F)$ is not in general a subgroup of $Z^n(K) + C_2^{n+1}(A,F)$ and it again is convenient to define the relative cohomology in terms of the subgroup, $\overline{Z}^n(A,F)$ generated by $Z^n(A,F)$. It is clear that $\overline{Z}^n(A,F) = \overline{Z}^n(A) + C_2^{n+1}(A,F)$.

Let $c_i = z_i + r_i \in \overline{Z}^n(A,F)$ $(i = 1,2)$. Then set $c_1 \equiv c_2(S)$ if $\overline{Z}(F) \oplus z_1 \oplus B^n(A) = \overline{Z}^n(F) \oplus z_2 \oplus B^n(A)$ and $dr_1 \oplus B^{n+1}(F) = dr_2 \oplus B^{n+1}(F)$. Finally set $H^n(A,F) = H^n(A,F;K) = \overline{Z}^n(A,F)/S$.

Proposition 3.3. *Let* F *be a* T-*subalgebra of* A. *Then the sequence* $\ldots \to H^n(F) \xrightarrow{\ell_n} H^n(A) \xrightarrow{j_n} H^n(A,F) \xrightarrow{k_n} H^{n+1}(F) \to \ldots$ *is exact, where* ℓ_n *and* j_n *are induced by the inclusion maps* $\overline{Z}^n(F) \to \overline{Z}^n(A)$, $\overline{Z}^n(A) \to \overline{Z}^n(A,F)$ *respectively and* k_n *is induced by the map* $z+r \to d_1 r\colon \overline{Z}^n(A,F) \to \overline{Z}^{n+1}(F)$.

Remarks 3.4. 1. Of course when K is not abelian the various elements of the above sequence are not groups. However, they all have zero elements and "exactness" is to be taken as meaning that the inverse image of zero under one map is equal to the image of its predecessor.

2. When K is abelian the usual exact sequence is obtained from the exact sequence $0 \to C^*(F) \to C^*(A) \to C^*(A)/C^*(F) \to 0$ of cochain complexes (see [4]), where $C^n(A) = \{c \mid c \in C_1^n,\ \text{al}(c) \subset A\}$. As we have already observed in this case $Z^n(C^*(A)) = \overline{Z}^n(A)$, $B^n(C^*(A)) = B^n(A)$ and $H^n(C^*(A)) = H^n(A)$. In general $Z^n(C^*(A)/C^*(F)) \neq \overline{Z}^n(A,F)$. However, the map $c \to C^n(F) + c : \overline{Z}^n(A,F) \to Z^n(C^*(A)/C^*(F))$ induces a map $\varphi_n\colon H^n(A,F) \to H^n(C^*(A)/C^*(F))$ such that

$$\begin{array}{ccccc}
H^n(A) & \longrightarrow & H^n(A,F) & \longrightarrow & H^{n+1}(F) \\
\downarrow & & \downarrow{\scriptstyle \varphi_n} & & \downarrow \\
H^n(C^*(A)) & \longrightarrow & H^n(C^*(A)/C^*(F)) & \longrightarrow & H^{n+1}(C^*(F))
\end{array}$$

is commutative. The five lemma now shows that the relative cohomology groups coincide.

The last topic that I wish to consider is the derivation of an exact cohomology sequence from an exact sequence

(1). $0 \to J \xrightarrow{\varphi} K \xrightarrow{\psi} L \to 0$ of compact abelian topological groups. (Here φ and ψ are continuous homomorphisms.)

Let A be a T-subalgebra of $\mathfrak{A}(u)$. Then it is clear how to define the maps $H^n(A;J) \to H^n(A;K) \to H^n(A;L)$. The problem is to define the connecting homomorphism K: $H^n(A;L) \to H^{n+1}(A;J)$.

The usual method is to take $z \in Z^n(A;L)$, find $c \in C^n(A;K)$ with $\psi \circ c = z$, $r \in C^{n+1}(A;J)$ with $\varphi \circ r = dc$ and to set $k[z] = [r]$. (Here [ ] denotes cohomology class.)

The problem is that it is not clear that c and r exist. Of course, since (1) is exact we can find $c \in C^n(K)$ and $r \in C^{n+1}(L)$ with the desired properties; but $al(c)$ and $al(r)$ need not be in A. What has to be shown is that the maps $x \to c(x,t_1,\ldots,t_n) : |A| \to K$ and $y \to r(y,t_1,\ldots,t_{n+1}) : |A| \to L$ are continuous $(t_1,\ldots,t_{n+1} \in T)$.

Lemma 3.5. With the above notation assume $al(c) \subset A$. Then $al(r) \subset A$.

Proof. By (1) of 2.5 $al(dc) \subset A$. Thus dc induces a continuous map f: $|A| \times T^{n+1} \to K$ such that $f(x|A,t) = dc(x,t)$ $(x \in \beta T,\ t \in T^{n+1})$.

Since $\psi \circ dc = 0$, there exists g: $|A| \times T^{n+1} \to J$ such that the diagram

$$\begin{array}{ccccc} 0 & \longrightarrow & J & \xrightarrow{\varphi} & K \\ & & \uparrow g & \nearrow f & \\ & & |A| \times T^{n+1} & & \end{array}$$

is commutative.

Let $x_i \to x$ in $|A| \times T^{n+1}$ and $g(x_i) \to y \in J$. Then $\varphi(y) = \lim g(x_i) = \lim f(x_i) = f(x) = \varphi(g(x))$, whence $y = g(x)$ since φ is injective. Thus g is continuous.

Now set $h(x,t) = g(x|A,t)$ $(x \in \beta T,\ t \in T^{n+1})$. Then h is continuous and $\varphi \circ h = dc = \varphi \circ r$ implies $h = r$. The proof is completed.

In order to be able to pick c with $al(c) \subset A$, I assume that $|A|$ is 0-dimensional and that J is a Lie group. The latter implies by a well known result of Gleason's that ψ admits local sections.

Lemma 3.6. <u>Let $|A|$ be 0-dimensional, J a Lie group, and $f: |A| \times T^n \to L$ continuous. Then there exists a continuous map $g: |A| \times T^n \to K$ with $f = \psi \circ g$.</u>

<u>Proof</u>. Fix $t \in T^n$. For every $x \in |A|$ there exist an open-closed neighborhood V_x of x and a continuous map $h_x: f_t(V_x) \to K$ ($f_t(y) = f(y,t)$ $(y \in |A|)$) such that $\psi \circ h_x$ is the identity on $f_t(V_x)$. Set $h_x^t(y) = h_x(f_t(y))$ $(y \in V_x)$. Then h_x^t is continuous on V_x and $\psi \circ h_x^t = f_t$ on V_x. Let $|A| = \cup \{V_{x_i} \mid 1 \le i \le n\}$, $j(x) = \min\{i \mid x \in V_{x_i}\}$, and $g_t(x) = h_{j(x)}^t(x)$ $(x \in |A|)$. Then g_t is continuous and $f_t = \psi \circ g_t$. The function $g(x,t) = g_t(x)$ satisfies the requirements of 3.6.

Corollary 3.7. <u>Let J be a Lie group $|A|$ 0-dimensional and $z \in Z^n(A,L)$. Then there exists $c \in C^n(K)$ with $dc = z$ and</u> $al(c) \subset A$.

Thus when $|A|$ is 0-dimensional and J is a Lie group the procedure discussed above may be used to define the connecting homomorphism. (That k is well defined is proved in the usual way.)

Proposition 3.8. <u>Let $0 \to J \to K \to L \to 0$ be an exact sequence of compact abelian groups with J a torus, and let $|A|$ be 0-dimensional. Then there exists an exact sequence</u>

$$\ldots \to H^n(A;J) \to H^n(A;K) \to H^n(A;L) \to H^{n+1}(A,J) \to \ldots .$$

<u>Remark 3.9</u>. When J, K, and L are not abelian one cannot use the

standard method to define the connecting homomorphism k. The problem is not so much in defining the image [r] of k[z] but rather in showing that al(r) ⊂ A.

REFERENCES

1. R. Ellis, Cocycles in topological dynamics, Topology (to appear).
2. R. Ellis, Lectures on Topological Dynamics, W.A. Benjamin Inc., New York, 1969.
3. S. MacLane, Homology, Academic Press Inc., New York, 1963.
4. K.E. Peterson, Extensions of minimal transformation groups, Math. Sys. Theory, Vol. 5, 365-375.

UNIVERSITY OF MINNESOTA

THE STRUCTURE OF SMALE DIFFEOMORPHISMS

by

John Franks

Among structurally stable diffeomorphisms the simplest and best understood are the so-called Morse-Smale diffeomorphisms which have a finite non-wandering set. In this article I want to survey what is known about a larger class of diffeomorphisms which have all the stability properties of the Morse-Smale diffeomorphisms but can have a much more complicated dynamics since we only assume the non-wandering set is zero dimensional instead of finite.

These diffeomorphisms satisfy Axiom A of Smale [8] which we now briefly describe.

Let $f: M \to M$ be a C^1 diffeomorphism of a compact connected manifold M. A closed f-invariant set $\Lambda \subset M$ is called <u>hyperbolic</u> if the tangent bundle of M restricted to Λ is the Whitney sum of two Df-invariant bundles, $T_\Lambda M = E^u(\Lambda) \oplus E^s(\Lambda)$, and if there are constants $C > 0$ and $0 < \lambda < 1$ such that

$$|Df^n(v)| \leq C\lambda^n|v| \qquad \text{for } v \in E^s, \ n > 0$$

and

$$|Df^{-n}(v)| \leq C\lambda^n|v| \qquad \text{for } v \in E^u, \ n > 0.$$

The diffeomorphism f is said to satisfy <u>Axiom A</u> if a) the non-wandering set of f, $\Omega(f) = \{x \in M : U \cap \bigcup_{n>0} f^n(U) \neq \emptyset$ for every neighborhood U of $x\}$ is a hyperbolic set, and b) $\Omega(f)$ equals the closure of the set of periodic points of f.

The <u>stable manifold</u> $W^s(x)$ of a point $x \in M$ is defined to be $\{y \mid d(f^n x, f^n y) \to 0$ as $n \to \infty\}$ and is an injectively immersed plane for each x if f satisfies Axiom A [8]. The unstable manifold $W^u(x)$ of x is simply the stable manifold of x for the diffeomorphism f^{-1}.

<u>Definition</u>. A diffeomorphism $f: M \to M$ is called a <u>Smale diffeomorphism</u> [11], [2], provided

1) It satisfies Axiom A;

2) For every $x \in M$, $W^s(x)$ and $W^u(x)$ intersect transversely.

3) The dimension of the non-wandering set $\Omega(f)$ is zero.

It was shown by Smale [9] that every isotopy class admits a Smale diffeomorphism and by Shub [5] that Smale diffeomorphisms are dense in the C^0 topology on Diff(M). If f satisfies Axiom A, one has the spectral decomposition theorem of Smale [8] which says $\Omega(f) = \Lambda_1 \cup \ldots \cup \Lambda_\ell$ where the Λ_i are pairwise disjoint, f-invariant closed sets and $f|_{\Lambda_i}$ is topologically transitive.

These Λ_i are called the <u>basic sets</u> of f and because f is topologically transitive on each basic set, the restrictions of the bundles E^s and E^u to Λ_i have constant dimension. The fiber dimension of $E^u(\Lambda_i)$ is called the <u>index</u> of Λ_i.

A result of Bowen [1] shows that on any zero-dimensional basic set Λ, f is topologically conjugate to a subshift of finite type, so we now describe these homeomorphisms.

Let C be a finite set and $r, \ell: C \to \{1,\ldots,n\}$ two functions. We give C the discrete topology and let

$$\Sigma = \{\underline{c} \in \prod_{-\infty}^{\infty} C \mid \ell(c_{i+1}) = r(c_i) \text{ for all } i \in Z\}.$$

Here $\underline{c}$ denotes the sequence $(c_i)_{i\in Z}$. The homeomorphism $\sigma: \Sigma \to \Sigma$, given by $\sigma(\underline{a}) = \underline{b}$ if $a_{i+1} = b_i$ for all i, is called a subshift of finite type.

The finite set and the functions can be completely described by an $n \times n$ non-negative matrix as follows: define A_{ij} to be the cardinality of $\{c \in C \mid \ell(c) = i, \ r(c) = j\}$. It is then clear that the matrix A completely determines the subshift of finite type $\sigma: \Sigma \to \Sigma$ and we will often denote it $\sigma: \Sigma_A \to \Sigma_A$. (We remark that there is another, different description of subshifts in terms of $0-1$ matrices (see [6] for example), but the one given here is better for our purposes.)

It is often useful to view subshifts in terms of paths on an oriented graph. Thus we can form a graph with n vertices labelled 1 to n and add A_{ij} oriented edges going from vertex i to vertex j, each of unit length. Then let S be the set of arclength and orientation preserving maps from the reals to this graph which map integers to vertices. If we give S the compact open topology and define $s: S \to S$ by $s(g)(t) = g(t+1)$, this is just another description of the subshift of finite type whose finite set C is the set of edges of the graph and with the functions $r, \ell: C \to \{1,\ldots,n\}$ assigning the right and left endpoint respectively.

Proposition 1.3 (Williams [10]). _If_ $A = RS$ _and_ $B = SR$, _and_ R, S _are non-negative integral matrices (not necessarily square) then the shifts_ $\sigma_A: \Sigma_A \to \Sigma_A$ _and_ $\sigma_B: \Sigma_B \to \Sigma_B$ _are topologically conjugate_.

In fact Williams [10] considers the equivalence relation created by making the relation $RS \sim SR$ transitive, which he calls _strong shift equivalence_, and shows that two matrices represent topologically conjugate subshifts if and only if they are strong shift equivalent. The matrices representing subshifts conjugate to Axiom A basic sets, can be chosen (see [1]) to be _irreducible_, i.e., with the property tnat for each ij, there is a $k > 0$, with $(A^k)_{ij} > 0$.

We now consider an example to illustrate how subshifts of finite type occur as basic sets. Figure 1 shows an embedding of X, the region bounded by a surface of genus 2, into itself. If we think of S^3 as R^3 plus a point at infinity it is not difficult to show this embedding can be extended to a diffeomorphism of S^3 with 3 fixed points (source and two saddles) being the only non-wandering points outside of X, but we wish to concentrate on the part of the

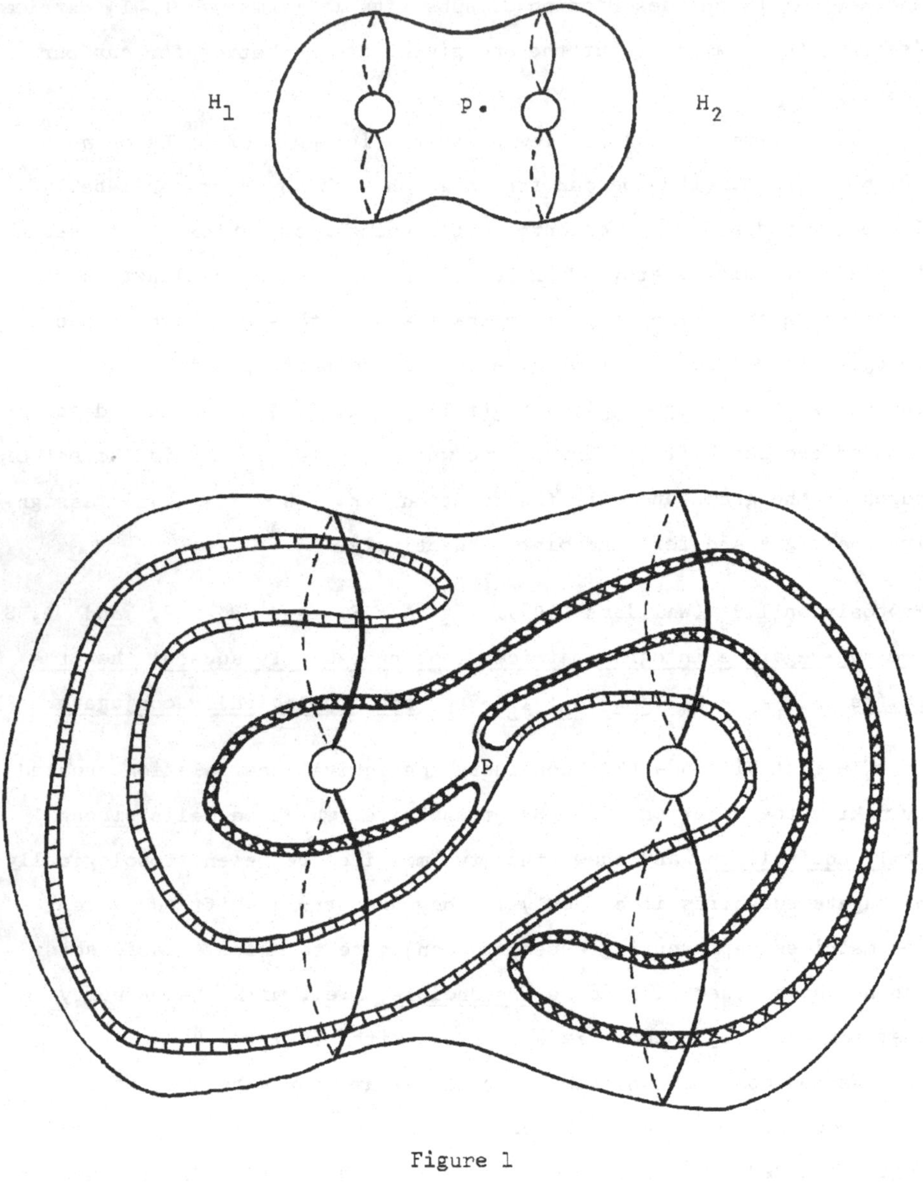

Figure 1

non-wandering set Ω which is inside of X. There is an attracting fixed point p and every other point in $X - \{H_1 \cup H_2\}$ tends toward p under iteration. Thus any remaining non-wandering points must have their entire orbit in $H_1 \cup H_2$.

In order to consider the set Λ of points whose entire orbits remain in $H_1 \cup H_2$ we define C = the set of components of $f(H_1 \cup H_2) \cap (H_1 \cup H_2)$. We can then define $h: \Lambda \to \prod_{-\infty}^{\infty} C$ by $h(x) = (\ldots c_{-1}, c_0, c_1, c_2, \ldots)$ where c_i is the element of C which contains $f^i(x)$. The map h is easily seen not to be surjective however since if $x \in c$ and c is a component of $H_1 \cap f(H_1)$ then $f(x)$ cannot be in $f(H_2)$ but must be in $f(H_1)$. Thus if c' is a component of $H_1 \cap f(H_2)$ we can never have c' follow c in any sequence which is $h(x)$ for some $x \in \Lambda$.

We can exclude these missed sequences by defining $r, \ell: C \to \{1,2\}$ by $\ell(c) = i$, $r(c) = j$ if c is a component of $H_i \cap f(H_j)$ and using these functions to define the subshift on $\Sigma \subset \prod_{-\infty}^{\infty} C$ as above. Note that the matrix corresponding to this subshift is

$$\begin{pmatrix} 1 & 2 \\ 2 & 1 \end{pmatrix}.$$

It is immediate from the definition that $h: \Lambda \to \Sigma$ is continuous and satisfies $\sigma \circ h = h \circ f$. If we have been careful in the construction of f, h will also be a homeomorphism. The care necessary roughly says that f is hyperbolic on $H_1 \cup H_2$, contracting into a two dimensional stable direction and expanding in a one-dimensional unstable direction, and that each $c \in C$ must extend all the way across (in the unstable direction) the H_i which contains it. Details can be found in §2 of [2]. These conditions are enough to insure however that for each $\underline{c} \in \Sigma$ the set $K_n = \bigcap_{i=-n}^{n} (c_i)$ is a non-empty compact set whose diameter tends to zero as $n \to \infty$. Thus $K_\infty = \bigcap_{-\infty}^{\infty} f^{-i}(c_i)$ consists of a single point and the map $\underline{c} \to K_\infty$ is an inverse to h.

Since h is a continuous bijection of compact spaces it is a homeomorphism.

We wish now to address the question of the relationship of the dynamics of a Smale diffeomorphism (its basic sets) to the topology of the manifold on which it occurs.

Since the direct sum of matrices corresponds to the disjoint union of the corresponding subshifts of finite type, we can describe all basic sets of a given index for a Smale diffeomorphism in a single matrix.

Definition. The direct sum of matrices corresponding to each of the basic sets of index j of a Smale diffeomorphism f will be called a jth type matrix for f (it is not unique).

Each of these matrices can be chosen to be a direct sum of irreducible matrices since the zero dimensional basic sets of an Axiom A diffeomorphism are conjugate to subshifts of finite type with irreducible matrices.

The following theorem of M. Shub and D. Sullivan ([4], [6]) shows the close relationship between the kinds of subshifts which can be realized as basic sets of a Smale diffeomorphism in a given isotopy class and homological invariants of that isotopy class.

Theorem (Shub, Sullivan). Let $f \in \mathrm{Diff}(M)$, $\Pi_1(M) = 0$, $\dim M = m \geq \sigma$. Suppose C is a finite free chain complex with $C_0 = C_m = Z$, $C_1 = C_{m-1} = 0$ and $E\colon C \to C$ is a chain endomorphism given as integer matrices $E_i\colon C_i \to C_i$. If $E_*\colon H_*(C;Z) \to H_*(C;Z)$ is the same as $f_*\colon H_*(M;Z) \to H_*(M;Z)$ and the same is true for Z/nZ coefficients for all n, then f is isotopic to a Smale diffeomorphism g with j^{th} type matrix $|E_i|$.

If A is an integer matrix $|A|$, of course, denotes the matrix whose ij^{th} entry is $|a_{ij}|$. If one is given non-negative integer matrices A_i it may be quite difficult to use this theorem to tell if a given isotopy class contains a Smale diffeomorphism with the A_i as type matrices since there is no easy way to check the existence of a chain map E with each $|E_i|$ shift equivalence to A_i.

However if we are willing to construct a Smale diffeomorphism with basic sets prescribed only up to a finite power then there are complete and simple conditions for realizability in the isotopy class of the identity. The answer depends on the reduced degree of the desired type matrices which we define as follows: the reduced degree of an integer matrix A is the degree of the mod 2 reduction of the polynomial $\det(I - At)$.

Theorem [2]. If M admits a Morse function whose type numbers are equal to its mod 2 Betti numbers, then a necessary and sufficient condition for the existence of a Smale diffeomorphism f on M, isotopic to the identity and with type matrices $\{A_j^k\}$, for some $k > 1$, is that

$$d_q - d_{q-1} + \ldots \pm d_0 \geq \beta_q - \beta_{q-1} + \ldots \pm \beta_0 \quad \text{for all } q,$$

and $\sum_{j=0}^{m} (-1)^j d_j = \chi(M)$, the Euler characteristic, where d_j is the reduced degree of A_j and β_j is the jth mod 2 Betti number of M. We assume each of the A_j is a direct sum of irreducible matrices, with A_0 and A_m permutation matrices, and that $\dim M > 2$.

By the jth mod 2 Betti number we mean the rank of $H_j(M; Z/2Z)$. Applying a result of Smale [7] one sees that the condition that M admit a Morse function with type numbers equal to its mod 2 Betti numbers is satisfied by any simply connected manifold of dimension greater than five if all of the torsion in homology is the direct sum of cyclic groups of even order.

The conditions of this Theorem are in fact necessary for the existence of a Smale diffeomorphism homotopic to the identity (not only isotopic) having the required type matrices.

Recently Steve Batterson has shown that this theorem also holds in the case that $\dim M = 2$.

We give a sketch of the main ideas of the proof of this theorem (details can be found in [2]). First one shows that there is a k such that A_i^k is shift equivalent to a matrix with all off-diagonal elements even. This is somewhat technical in general but in the case of matrices with odd determinant it is clear since the mod 2 reduction of the matrices will be elements of finite order in $Gl(n,Z/2)$. It is these matrices with even off-diagonal entries which we realize as type matrices. Note the number of odd (diagonal) entries is the reduced degree of the matrix (before or after it is raised to a power). The construction begins by constructing a Morse function with exactly d_i critical points of index i. The inequalities guarantee that this is possible. As a first approximation to the desired diffeomorphism we take g, the time one map of the flow of minus the gradient of this Morse function. Corresponding to this Morse function there is a handle decomposition preserved by g. The type matrices for g are diagonal $0-1$ matrices, in fact exactly the mod 2 reduction of the desired matrices. By repeated isotopies of the type illustrated in Figure 2 one can change the type matrices by adding any even number to any entry and thus the desired type matrix can be achieved. The necessity of the conditions follows from results of [3].

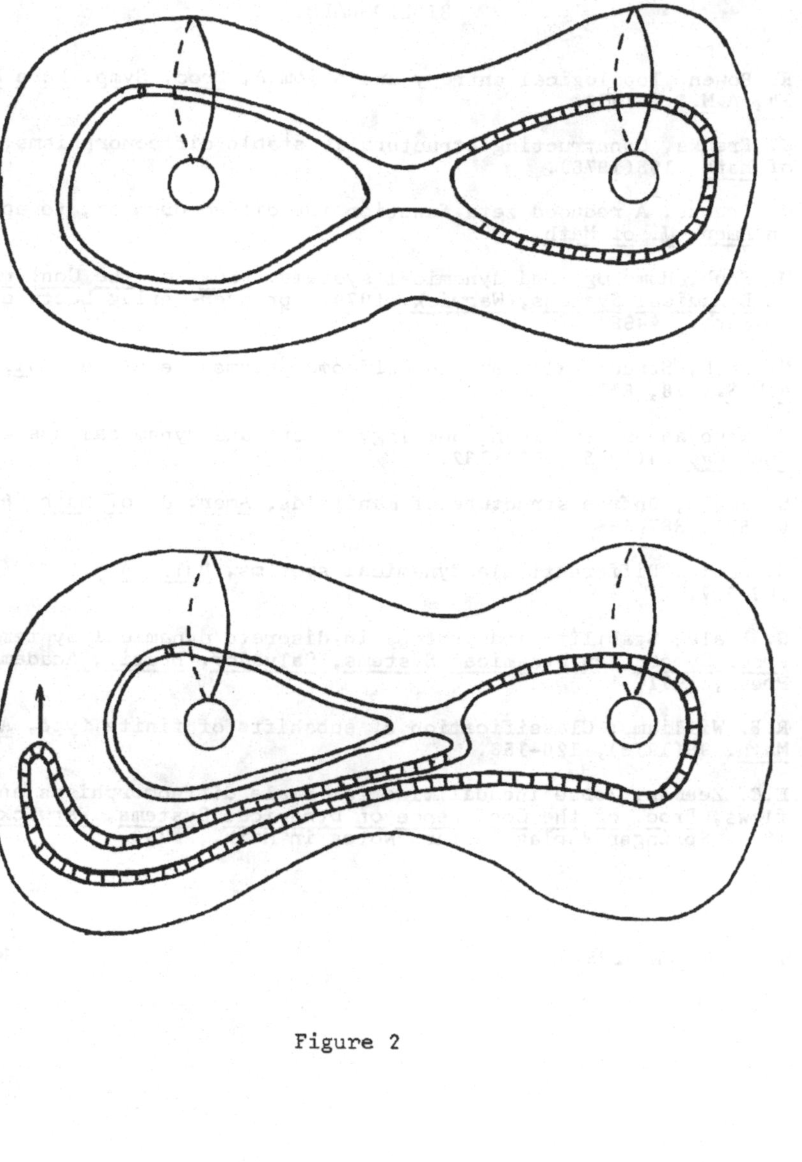

Figure 2

BIBLIOGRAPHY

1. R. Bowen, Topological entropy and Axiom A, Proc. Symp. Pure Math. 14, A.M.S., 1970.

2. J. Franks, Constructing structurally stable diffeomorphisms, Ann. of Math. 105(1976).

3. J. Franks, A reduced zeta function for diffeomorphism, to appear in Amer. J. of Math.

4. M. Shub, Homology and dynamical systems, Proc. of the Conference on Dynamical Systems, Warwick, 1974, Springer-Verlag Lecture Notes in Math., #468.

5. M. Shub, Structurally stable diffeomorphisms are dense, Bull. A.M.S., 78, 817.

6. M. Shub and D. Sullivan, Homology theory and dynamical systems, Topology 14(1975), 109-132.

7. S. Smale, On the structure of manifolds, Amer. J. of Math. 84 (1962), 387-399.

8. S. Smale, Differentiable dynamical systems, Bull. A.M.S. 73(1967), 747-817.

9. S. Smale, Stability and isotopy in discrete dynamical systems, Proc. Sympos. on Dynamical Systems, Salvador, Brazil, Academic Press, 1971.

10. R.F. Williams, Classification of subshifts of finite type, Ann. of Math. 98(1973), 120-153.

11. E.C. Zeeman, Morse inequalities for Smale diffeomorphisms and flows, Proc. of the Conference on Dynamical Systems, Warwick, 1974, Springer-Verlag Lecture Notes in Math., #468.

NORTHWESTERN UNIVERSITY

THE FINITE MULTIPLIERS OF INFINITE ERGODIC TRANSFORMATIONS

by

Hillel Furstenberg and Benjamin Weiss

It is well known that a finite measure preserving transformation $(X,\mathcal{B},\mu,T)$, $(\mu(X)=1)$, satisfies

(A) <u>For all finite ergodic</u> $(Y,\mathcal{C},\nu,S)$ <u>the product</u> $(X\times Y,\mathcal{B}\times\mathcal{C},\mu\times\nu,T\times S)$ <u>is also ergodic</u>

if and only if T has no non-trivial point spectrum, i.e. is weakly mixing. The question we settle here is what happens if one drops the assumption that $\nu(Y)<\infty$ in (A). It is claimed in [2] that the answer remains the same, in other words that for T weakly mixing and S ergodic, even if $\nu(Y)=\infty$, $T\times S$ is ergodic. As we shall see below this is not the case, and one must broaden the notion of point spectrum if one is to characterize the class of <u>finite</u> transformations that satisfy:

(B) <u>For all ergodic</u> $(Y,\mathcal{C},\nu,S)$ <u>the product</u> $(X\times Y,\mathcal{B}\times\mathcal{C},\mu\times\nu,T\times S)$ <u>is also ergodic</u>.*

If T has non-trivial spectrum then for some non-constant function f, and some sequence n_i, $T^{n_i}f\to f$ in L_2. Indeed if $Tf=\lambda f$ with $|\lambda|=1$ then there certainly are sequences n_i such that $\lambda^{n_i}\to 1$. This is the relevant feature of the existence of eigenvalues to determine what transformations T will satisfy (B). Formally, let us say that T is <u>rigid</u> if there exists a sequence $n_i\uparrow\infty$ such that $T^{n_i}f\to f$ in L_2 for all $f\in L_2$. This notion is already implicit in the work of S. Foguel [3]. Our main result is:

*(Throughout this discussion ergodic will mean ergodic and conservative; i.e. we exclude the trivial transformation: $Y=\mathbb{Z}$, $Sn=n+1$.)

Theorem. <u>A transformation</u> $(X,\mathcal{B},\mu,T)$ <u>satisfies</u> (B) <u>if and only if</u> T <u>has no non-trivial rigid factor</u>.

It is not very difficult to construct examples of rigid transformations that are weakly mixing. Either the method of cutting and stacking, or starting with a continuous measure σ on the circle such that $\overline{\lim_{n\to\infty}} |\hat{\sigma}(n)| = 1$, and then constructing a Gaussian process with σ as covariant function will work. On the other hand, it is clear that if T is strongly mixing then T has no rigid factors. Finally, it is also not difficult to see that there are transformations with no rigid factors that are not strongly mixing. Again one either uses the method of cutting and stacking or prescribes spectral measures.

Thus the class of transformations with no rigid factors lies strictly between the strongly mixing and the weakly mixing transformations, and might be called <u>mildly mixing</u>. We don't propose to study this class in detail here, but mention one trivial consequence of the theorem -- namely that the product of mildly mixing transformations is again mildly mixing.

We will prove the theorem contrapositively, separating out the two parts.

Proposition 1. <u>If</u> T <u>is rigid there is an ergodic</u> $(Y,\mathcal{C},\nu,S)$ <u>such that</u> $T \times S$ <u>is not ergodic</u>.

<u>Proof</u>. Fix some non-constant function f with zero integral and norm 1 in L_2, and then find a sequence $\{q_i\}_1^\infty$ such that for all finite sets $\alpha \subset N = \{1,2,3,\ldots\}$

$$\|T^{q_\alpha} f - f\|_2 \le \frac{1}{10}$$

where $q_\alpha = \sum_{i\in\alpha} q_i$. To find the q_i proceed inductively, letting q_1 be such that $\|T^{q_1} f - f\|_2 \le \frac{1}{10}$ and then choosing q_2 so that $T^{q_2} f$ is close enough to f to ensure that both $\|T^{q_2} f - f\|_2 \le \frac{1}{1000}$ and also

$\|T^{q_1+q_2}f - f\|_2 \le \frac{1}{100} + \frac{1}{1000}$, and so on.

Next let $\Omega = \{0,1\}^N$ and let $\tau: \Omega \to \Omega$ be addition by 1 with carry to the right, i.e.

$$\tau(\omega_1,\omega_2,\ldots) = (0,0,0,\ldots,0,1,\omega_{n+1},\ldots)$$

if $\omega_1 = \omega_2 = \ldots = \omega_{n-1} = 1$, but $\omega_n = 0$.

$$\tau(0,\omega_2,\omega_3,\ldots) = (1,\omega_2,\omega_3,\ldots)$$

and

$$\tau(1,1,1,\ldots) = (0,0,0,\ldots).$$

Then τ preserves the product measure $(1/2,1/2)^N$ and is known to be ergodic. Define now an integer valued function $h: \Omega \to N$ by

$$h(0,\omega_2,\ldots) = q_1$$

$$h(1,1,\ldots,1,0,\omega_{n+1},\ldots) = q_n - (q_1 + q_2 + \ldots + q_{n-1})$$

while $h(1,1,\ldots,1,1,\ldots) = 1$.

Finally let (Y,S) be the tower built over (Ω,τ) with height function h, i.e.

$$Y = \{(\omega,j): \ \omega \in \Omega, \ 1 \le j \le h(\omega)\}$$

$$S(\omega,j) = \begin{cases} (\omega,j+1) & \text{if } j+1 \le h(\omega) \\ (\tau\omega,1) & \text{if } j = h(\omega) \end{cases}$$

and let (Y,C,ν,S) be the process generated in this way. Since τ was ergodic so is S. If $T \times S$ were to be ergodic then the transformation induced by $T \times S$ on the set

$$A = \{(x,\omega,1) : \ x \in X, \ \ \omega \in \Omega\}$$

would also be ergodic. Identifying A with $\Omega \times X$, and letting U denote the induced transformation we see that U is given by

$$U(\omega,x) = (\tau\omega, T^{h(\omega)}x).$$

Now defining h_n recursively by, $h_1 = h$ and

$$h_{n+1}(\omega) = h_n(\omega) + h_1(\tau^n\omega)$$

we have that

$$U^n(\omega,x) = (\tau^n\omega, T^{h_n(\omega)}x).$$

The crucial observation is, that with h as we have defined it,

$$h_n(\omega) \in \{q_\alpha - q_\beta : \alpha,\beta \subset N\}$$

but for countably many values of ω. Furthermore, since T is unitary on L_2, from $\|T^{q_\alpha}f - f\|_2 \leq \frac{1}{10}$ for all finite $\alpha \subset N$, follows also

$$\|T^{q_\alpha - q_\beta}f - f\|_2 \leq \frac{1}{5}$$

for all finite subsets α, β. Using f to define a function $\tilde{f}$ on $\Omega \times X$ by $\tilde{f}(\omega,x) = f(x)$ it is now completely clear that

$$\frac{1}{N}\sum_{1}^{N} U^n\tilde{f}$$

cannot converge to the integral of $\tilde{f}$, and thus U is not ergodic.

For the other direction we prove now

Proposition 2: <u>If for an ergodic</u> (Y,C,ν,S), $T \times S$ <u>is not ergodic then</u> T <u>has a rigid factor</u>.

<u>Proof</u>. Let $A \subset X \times Y$ be a non-trivial $T \times S$ invariant set. Denote by $f_y(x) = 1_A(x,y)$. Since A is non-trivial while

$$\int f_y(x)\, d\mu(x)$$

is an S-invariant function we must have that for some $0 < c < 1$ and almost all y,

$$\int f_y(x)\, d\mu(x) = c.$$

The invariance of A leads to the equation

$$f_{Sy}(Tx) = f_y(x). \qquad (*)$$

We will obtain now the fact that there are y_0, with $\int f_{y_0}(x)\, d\mu(x) = c$, and a sequence $n_i \uparrow \infty$ with $T^{n_i} f_{y_0} \to f_{y_0}$ by using simply the recurrence properties of S and the separability of $L_2(X)$. A similar argument occurs in [1]. It follows from Fubini's theorem that the mapping from Y to $L_2(X)$ given by $y \to f_y$ is weakly measurable, and thus by Pettis' theorem it's strongly measurable. We can therefore determine an increasing sequence of countable partitions of Y, $\ldots\ \alpha_1 \subset \alpha_2 \subset \ldots$ such that if y, y' are in the same set of the kth partition α_k we have

$$\|f_y - f_{y'}\|_2 \le 10^{-k}.$$

Apply Poincaré recurrence to every set of each α_k using the conservativeness of (Y,C,ν,S), and collecting the countable collection of null sets together, we have that for almost every y, there is a sequence n_k such that

$$\left\|f_y - f_{S^{n_k}y}\right\|_2 \le 10^{-k}.$$

Recalling (*) we have that, as promised, $T^{n_k} f_y \to f_y$ with f_y non-trivial. For a fixed sequence n_k, the set of g that satisfy $T^{n_k} g \to g$ is a closed algebra and thus T has a non-trivial rigid factor.

It is straightforward to check that σ is the spectral measure of a mildly mixing transformation if and only if

$$\overline{\lim_{|n|\to\infty}}\ |\hat{\sigma}(n)| < \hat{\sigma}(0). \qquad (*)$$

Measures that satisfy (*) have been recently considered by C. Moore and K. Schmidt in connection with some questions in the theory of co-cycles. They call such measures "full" measures. The construction that we give in the proof of Proposition 1 also serves to give an example in that theory as follows: The function h that is constructed depends only on the sequence $q_1 < q_2 < \dots$. Since $h \geq 1$ it is clear that h is not a co-boundary, namely there is no solution to the additive equation

$$h(\omega) = a(\tau\omega) - a(\omega).$$

However, if the q_i's satisfy a suitable growth condition there are uncountably many values of λ, for which the equation

$$\exp 2\pi i \lambda h(\omega) = \alpha_\lambda(\tau\omega)/\alpha_\lambda(\omega) \qquad (**)$$

has a solution with $\alpha\colon \Omega \to T^1$. Namely let

$$\Lambda = \{\lambda : \sum_1^\infty |1 - \exp 2\pi i \lambda q_n| < \infty\},$$

and then set

$$\alpha_\lambda(\omega) = \prod_1^\infty \exp 2\pi i \lambda q_n \omega_n$$

and verify that (**) indeed holds.

REFERENCES

1. Anatole Beck, Eigen operators of ergodic transformations, Trans. Amer. Math. Soc. 94(1960).

2. Elias G. Flytzannis, Ergodicity of the Cartesian product, Trans. Amer. Math. Soc. 186(1973), 171-176.

3. Shaul Foguel, Invariant subspaces of a measure preserving transformation, Israel J. of Math. 2(1964), 198-200.

HEBREW UNIVERSITY

APPLICATIONS OF ERGODIC THEORY TO GEOMETRY

by

Leon W. Green[1]

This talk is in the nature of propaganda for a technique, namely the use of ergodic theory in global differential geometry. I will review some notable applications (and "near" applications -- see Section 2) and indicate a few recent results. But in my opinion the possibilities have barely been touched, so this is an invitation to those of you who are geometrically inclined to join the game.

All the applications I have in mind utilize the geodesic flow or flows closely related to it; let me recall what it is:

Let M be a complete, connected smooth Riemannian manifold. T_1M its unit tangent bundle, π the projection of T_1M onto M. If $\gamma: \mathbb{R} \to M$ is a geodesic parameterized by arc-length with initial condition $\dot{\gamma}(0) = v$ for $v \in T_1M$, then by definition, $f_t v = \dot{\gamma}(t)$. This defines an action of $\mathbb{R}$ on T_1M called the geodesic flow. X will denote the vector field of this flow. The Riemannian structure induces a volume element Ω in T_1M and $\{f_t\}$ preserves the resulting measure.

§1. Two Applications

Application 1: Mostow's Rigidity Theorem. In 1973 G.D. Mostow published his book "Strong Rigidity of Locally Symmetric Spaces," [10]. His main result is that, modulo easily described exceptions, a compact, connected, locally symmetric Riemannian manifold of non-positive curvature is essentially determined by its fundamental group. Here "essentially determined" means "within an isometry and choice of normalizing constants." The details of the proof of this deep theorem occupy considerably more than the cited book. (For an outline of the proof, see

[1]Supported in part by National Science Foundation (MCS77-9747).

Koszul [7].) Here I merely want to point out that, at a crucial step in the rank one case, Mostow appeals to the ergodicity of the geodesic flow in such spaces, a result obtained in 1957 by Mautner [9]. This cannot be called the heart of the proof, but it certainly seems essential. Mostow's theorem is the outstanding example of a differential geometry result which uses ergodic theory.

Application 2: Spaces without conjugate or focal points. A connected, complete Riemannian manifold is said to have no conjugate points if, in its universal covering space, any two points are joined by one and only one geodesic. Spaces covered by Euclidean or hyperbolic spaces are prime examples. The circle of ideas I have in mind under application 2 revolves around a theorem of E. Hopf: A compact orientable surface with no conjugate points has non-positive total curvature; if it is homeomorphic to a torus, it necessarily has zero Gaussian curvature [5]. Hopf's proof is elegant and does not use any ergodic theory ideas beyond the invariance of Ω under the geodesic flow. Nevertheless, anticipating later sections of this talk, I shall present a proof which invokes the Birkhoff ergodic theorem.

If M is a compact surface without conjugate points, with Gaussian curvature K, it is possible to construct a bounded, continuous real-valued function u on T_1M which satisfies the Riccati equation

$$(*) \qquad Xu + u^2 + K\circ\pi = 0.$$

Geometrically, $u(v)$ represents the geodesic curvature of the limit circle, or horocycle, with exterior normal v in the covering surface of M. Set $v_t = f_t v$ and calculate the time averages along the geodesic flow of the terms in equation (*):

$$\frac{1}{T}\int_0^T (Xu)(v_t)\,dt + \frac{1}{T}\int_0^T u^2(v_t)\,dt + \frac{1}{T}\int_0^T (K\circ\pi)(v_t)\,dt = 0.$$

Since u is bounded,

$$\lim_{T\to\infty} \frac{1}{T}\int_0^T (Xu)(v_t)\,dt = \lim_{T\to\infty} \frac{1}{T}\{u(v_T) - u(v_0)\} = 0.$$

On the other hand, by the ergodic theorem there exist functions w^* and k^* such that, for almost all v,

$$\lim_{T\to\infty} \frac{1}{T}\int_0^T u^2(v_t)\,dt = w^*(v)$$

and

$$\lim_{T\to\infty} \frac{1}{T}\int_0^T (K\circ\pi)(v_t)\,dt = k^*(v).$$

Thus, equation (*) implies that for almost all v,

$$w^*(v) + k^*(v) = 0.$$

Consequently

$$\int w^*(v)\Omega + \int k^*(v)\Omega = 0.$$

But the Birkhoff ergodic theorem says that the space averages of these time-average functions equal the integrals of the original functions:

$$\int_{T_1M} w^*\Omega = \int_{T_1M} u^2\Omega, \qquad \int_{T_1M} k^*\Omega = \int_{T_1M} (K\circ\pi)\Omega;$$

so we conclude that

$$(**) \qquad \int_{T_1M} u^2\Omega + \int_{T_1M} (K\circ\pi)\Omega = 0.$$

Since

$$\int_{T_1M} (K\circ\pi)\Omega = \int_M\int_0^{2\pi} (K\circ\pi)\,d\varphi dA = 2\pi\int_M K\,dA,$$

equation (**) implies the first part of Hopf's theorem. In the torus case, the Gauss-Bonnet theorem requires the total curvature to be zero. Therefore, by (**), u^2 is zero almost everywhere and (*) then leads to the conclusion that $K = 0$.

In the final version of his proof, Hopf avoided the use of the ergodic theorem by integrating (*) for only a finite t-interval ([0,1] will do), then integrating with respect to Ω and using the invariance of Ω to arrive directly at (**). However, I have given this proof because it contains one of the basic ideas of a theorem of Avez' [1]: *If the compact, connected Riemannian manifold M has no focal points, then either $\pi_1(M)$ has exponential growth or M is flat.* (The condition "no focal points" here is equivalent to the convexity of spheres in the universal covering space.) Thus Avez' theorem is a partial generalization of Hopf's, since it implies that n-dimensional tori without focal points are flat. The technique is to use an analogue of equation (*) for spheres of finite radius and compare the growth rate of the volumes of these spheres with the sign restriction implied by a generalization of (**). In higher dimensions, Hopf's trick for eliminating the use of the ergodic theorem gives a much weaker result. (O'Sullivan [11] has obtained part of this theorem by a different method.)

§2. Other Applications

In recent years there has been a great deal of interest in what I call results of a mixed geometrical-dynamical type. By this I mean a theorem where the hypothesis is geometrical and the conclusion dynamical, or vice versa. For instance, Klingenberg has proved [6]: *If the geodesic flow is Anosov, then M has no conjugate points.* An example of a theorem in the other direction has been announced by Gromov [4]: *If $\pi_1(M)$ has k generators and one relation, then the topological entropy of the geodesic flow is greater than or equal to* log (k-1). (Here M is compact with diameter less than or equal to one.)

"Pure" applications should, however, have geometrical hypotheses and geometrical conclusions, as the examples mentioned in Section 1 do. Here is a simple example which uses a flow different from the geodesic flow. It may be called a generalization of an old result attributed to K. Brauner (Blaschke [2], problem 9, page 192.) For a surface M suppose there exists a constant $r > 0$ such that every circle of radius r has constant geodesic curvature. Make enough assumptions about M so that sufficiently many such circles exist. Then Brauner's result is that M has a constant Gaussian curvature. The generalization I have in mind is obtained by letting r go to infinity; that is, suppose M is now a complete surface without conjugate points. If every horocycle of M has constant geodesic curvature, then M has constant Gaussian curvature.

Proof # 1 (using dynamics): Suppose in addition that M is compact and that its geodesic flow is Anosov. Then the unstable foliation is minimal. When orientable, this may be phrased: the expanding horocycle flow is minimal. At any rate, at least one of its leaves is dense. But the geodesic curvature of horocycles may be considered a function on the unit tangent bundle, which, by hypothesis, is constant on each of these leaves. Therefore the geodesic curvatures of all (expanding) horocycles are equal, and equation (*) of Section 1 may be used to complete the proof.

Unfortunately for my theme, there is another proof which is more general, simpler, and does not use dynamics. Here it is:

Proof # 2: Choice of an orientation in the universal covering surface of M allows us to define the expanding horocycle flow vector field H. Its relationship to the geodesic flow is given by the equation

$$(***) \qquad [X,H] = -uH,$$

where $u(v)$ is the geodesic curvature of the horocycle determined by v. The hypothesis now reads $H(u) = 0$, so when we apply H to equation (*) and use the bracket relation, we find that

$$
\begin{aligned}
0 &= HX(u) + H(u^2) + H(K\circ\pi) \\
&= [H,X](u) + XH(u) + H(u^2) + H(K\circ\pi) \\
&= -uH(u) + XH(u) + H(u^2) + H(K\circ\pi) \\
&= H(K\circ\pi).
\end{aligned}
$$

Hence K is constant on horocycles. But any two points of M lie on at least one horocycle (only one, if M is simply connected), so K must be constant.

This proof does not use the compactness of M and the curvature restrictions can be relaxed a little. However, some hypothesis beyond the absence of conjugate points seems to be needed to insure the smoothness of u and H.

§3. More on Horocycles

I have now introduced enough notation to describe a geometric result whose proof is less trivial than the preceding one. But first let me give the dynamical background. Responding to a question raised by B. Marcus in [8], I was recently able to prove that a three-dimensional Anosov flow on a compact manifold which admits smooth uniformly expanding (respectively, contracting) transverse flows is really a flow induced by a one-parameter subgroup of a Lie group on a homogeneous space. (See [3] for details of this theorem and of the proofs for the rest of this section.) If the Anosov flow in question was the geodesic flow on a surface of negative curvature, this result says that the unit tangent bundle has a very nice structure, but leaves open the question of the Riemannian structure of the original surface. One would expect that a surface with such a tangent bundle would have constant curvature, and such is indeed the case. But the only proof I know involves ergodic theory, even though the geometrical result can be disguised to hide its dynamical origin. Here are some more precise statements and an outline of the proof.

The assumptions on the surface M imply that the expanding and contracting horocycle flows exist and their vector fields (designated by H^+ and H^-, respectively) satisfy relations like (***) with respect to the geodesic flow, X:

$$[X,H^+] = -uH^+, \qquad [X,H^-] = -\bar{u}H^-.$$

H^+ will be said to have a smooth uniformly expanding reparameterization if there exists a nowhere zero C' function g on T_1M such that the vector field defined by the equation

$$E^u = gH^+$$

satisfies the commutation relation

$$[X,E^u] = -\mu E^u,$$

for μ a positive constant. Then we can state the

Theorem: _If the compact surface of negative curvature M admits a smooth uniformly expanding reparameterization of its expanding horocycle flow, then the curvature of M is constant._

(i) The first step is to show that H^- admits a uniformly contracting reparameterization. In fact, because of equations g must satisfy, the function

$$h = [(u-\bar{u})g]^{-1}$$

is a uniform reparameterization function for H^-. Set $E^s = hH^-$.

(ii) By the result cited about this special kind of Anosov flow, X, E^u, and E^s span a Lie subalgebra of vector fields on T_1M isomorphic to $\mathfrak{sl}(2,\mathbb{R})$:

$$[X,E^u] = -\mu E^u, \qquad [X,E^s] = \mu E^s, \qquad [E^u,E^s] = aX.$$

But further geometric conditions which H^+ and H^- must satisfy lead to the important normalization $a = 1$. Thus there is a cocompact discrete subgroup Γ of the universal covering group G of $SL(2, \mathbb{R})$ and a diffeomorphism η of T_1M onto G/Γ which is equivariant with respect to the flows generated, respectively, by X on T_1M and a semi-simple element of $\mathfrak{sl}(2, \mathbb{R})$ on G/Γ.

(iii) Because of ergodicity, Ω on T_1M and the measure ν on G/Γ induced by Haar measure correspond under η. Then the "coefficient of expansion" μ may be identified with the entropy of the geodesic flow in T_1M with respect to Ω and, by the isomorphism of the corresponding flow on G/Γ with respect to ν.

(iv) The proof now consists in computing μ in terms of the function u in two different ways, using the two flows. On the one hand,

$$\int_{T_1M} u\Omega = \mu.$$

This can be computed directly from a differential equation satisfied by g, or it follows from more general results of Sinai ([12], Theorem 7.1). On the other hand, if $\chi(M)$ denotes the Euler characteristic of M,

$$\mu^2 = -4\pi^2\chi(M).$$

This equality comes from looking at the flow in G/Γ as if it came from a flow on a surface of constant negative curvature. But then equation (**) of Section 1 and the Gauss-Bonnet theorem lead to

$$\int_{T_1M} u^2\Omega = \mu^2.$$

Comparing these two expressions for μ in light of the Schwarz inequality, we find that the function u must be constant. As we noted several times before, this (via equation (*) of Section 1) leads directly to the conclusion that the curvature is constant.

§4. Conclusion

In conclusion, I would like to repeat that the traditional interplay between dynamics and differential geometry is by no means limited to Hamiltonian formalism. In particular, especially in the study of compact manifolds of negative curvature, I expect that the entropy of the geodesic flow will prove to be an interesting geometric invariant.

REFERENCES

1. A. Avez, Variétés riemanniennes sans points focaux, C.R. Acad, Sci. Paris, Ser. A-B, 270(1970), A188-A191.*

2. W. Blaschke, Vorlesungen über Differentialgeometrie I, 3rd Aufl., Berlin, Springer, 1930.

3. L.W. Green, Remarks on uniformly expanding horcycle parameterizations, J. Diff. Geom., to appear.

4. M. Gromov, Three remarks on geodesic dynamics and the fundamental group, preprint.

5. E. Hopf, Closed surfaces without conjugate points, Proc. Nat. Acad. Sci. USA, 34(1948), 47-51.

6. W. Klingenberg, Riemannian manifolds with geodesic flow of Anosov type, Annals of Math., 99(1974), 1-13.

7. J.-L. Koszul, Rigidité forte des espaces riemanniens localement symétriques, Séminaire Bourbaki, Lecture Notes in Mathematics, no. 514, Springer, 1976, Exp. 468.

8. B. Marcus, Ergodic properties of horocycle flows for surfaces of negative curvature, Annals of Math., 105(1977), 81-106.

9. F.I. Mautner, Geodesic flows on symmetric Riemann spaces, Annals of Math., 65(1957), 416-431.

10. G.D. Mostow, Strong rigidity of locally symmetric spaces, Ann. of Math., 78(1973).

11. J.J. O'Sullivan, Manifolds without conjugate points, Math. Ann. 210 (1974), 295-311.

12. J.G. Sinai, Classical dynamical systems without countable Lebesgue spectrum, II, Izvestia Math. Nauk, 30(1966), 15-68.

UNIVERSITY OF MINNESOTA

---

*See also:

G.A. Margulis, On some applications of ergodic theory to the study of manifolds of negative curvature, Funk. Anal. i Priložen, 3(1969), 89-90.

ON EXPANSIVE HOMEOMORPHISMS OF THE INFINITE TORUS

by

Harold M. Hastings*

We prove that there are no expansive linear homeomorphisms of the infinite torus.

§1. Introduction

A homeomorphism f of a metric space X is called <u>expansive</u> if there is an $\varepsilon > 0$, such that for any two distinct points x and y of X, $\mathrm{dist}(f^n x, f^n y) > \varepsilon$ for some integer n. Anosov diffeomorphisms [1] of compact metric spaces are clearly expansive, and any invertible matrix with integral entries and no eigenvalues of modulus one yields an Anosov diffeomorphism of the appropriate finite dimensional torus. For example, the matrix

$$\begin{pmatrix} 2 & 1 \\ 1 & 1 \end{pmatrix}$$

yields an Anosov diffeomorphism of $T^2 = R^2/Z^2$. All known Anosov diffeomorphisms arise in essentially this way (J. Franks [3], A. Manning [4]). We shall prove that there are no expansive linear homeomorphisms of R^∞ (R^∞ has the product topology), hence, no expansive linear homeomorphisms of T^∞.[1]

R.F. Williams suggested the possibility of a relationship between this study and recent theories of turbulence. We thank Williams and J. West for helpful conversations.

We review some of the algebra and topology of R^∞ and T^∞ in Section 2, and give the proof in Section 3.

*Partially supported by National Science Foundation (MCS77-01628).

[1]We have recently learned that R. Mané independently proved that there are no expansive homeomorphisms of T^∞.

§2. Background

The infinite torus $T^\infty = \prod_{i=1}^\infty S^1$ is the quotient of R^∞ by the natural action of Z^∞. Although the map $R^\infty \to T^\infty$ is not a covering map, it is the inverse limit of covering maps

$$\lim_n \{R^n \times \prod_{i=n+1}^\infty S^1 \longrightarrow T^n \times \prod_{i=n+1}^\infty S^1\},$$

hence an open principal fibration in the sense of J. Cohen [2]. We therefore regard R^∞ as a kind of "tangent space" to T^∞, and define a linear homeomorphism of T^∞ to be a homeomorphism induced by a linear homeomorphism of R^∞.

The structure of the product metric on R^∞ is central to the non-existence of expansive linear homeomorphisms of R^∞ (and hence of T^∞). Denote points in R^∞ by sequences, e.g.,

$$x = (x_1, x_2, \ldots).$$

Then

$$\operatorname{dist}(x,y) = \sum \frac{|x_n - y_n|}{2^n(1 + |x_n - y_n|)}.$$

Thus, for any $\varepsilon > 0$, there is an N such that $\operatorname{dist}(x,0) < \varepsilon$ whenever the first N coordinates of x are zero. For example, this implies that multiplication by 2 does not yield an expansive homeomorphism of R^∞. We shall write $|x|$ for $\operatorname{dist}(x,0)$ but caution that $|\;|$ is <u>not</u> a norm. However, at least, $|x+y| < |x| + |y|$.

Let

$$R_w^\infty = \cup R^n \subset R^\infty$$

be the (dense, non-closed) subspace of R^∞ consisting of sequences which are eventually 0. Let $(e_1, e_2, \ldots)$ be the usual ordered basis for R_w^∞; e_i has 1 for its ith coordinate, and zeroes elsewhere. Then $\{e_1, e_2, \ldots\}$ is a basis for R^∞ <u>as a topological vector space</u>:

any x in R can be written as $\sum x_n e_n$.

There is an inner product $< , >$ on R_w^∞ and a pairing $R_w^\infty \times R^\infty \to R$, also denoted $< , >$ which makes R_w^∞ and R^∞ duals of each other; in both cases,

$$<x,y> = \sum_{n=1}^{\infty} x_n y_n .$$

Thus any continuous linear map f from R^∞ to itself has an adjoint $g: R_w^\infty \to R_w^\infty$; $<x,fy> = <gx,y>$. If f is a homeomorphism, so is g. As usual, the matrix of g with respect to $(e_1,e_2,\ldots)$ is the transpose of the matrix of f with respect to the corresponding $(e_1,e_2,\ldots)$.

§3. Proof

Let f be a linear homeomorphism of R^∞, and let $\varepsilon > 0$. We shall construct a vector x in R^∞ with $|f^n x| < \varepsilon$ for all integers n.

Choose N so that $|x| < \varepsilon/2$ for all vectors x in R^∞ whose first N coordinates are 0. Let $g: R_w^\infty \to R_w^\infty$ be the adjoint of f. Let $\pi_i: R^\infty \to R$ be the projection onto the ith coordinate.

Proposition 3.1. <u>There is a vector</u> x <u>in</u> R^∞ <u>with</u>

$$|\pi_i f^n x| = |<e_i,f^n x>| = |<g^n e_i,x>| < \varepsilon/2N, \qquad 1 \le i \le N, \quad n \in Z.$$

<u>Proof</u>. Let $E_i = \text{span}\{g^n e_i\}$ for $1 \le i \le N$ (<u>unless otherwise stated</u>, n <u>always ranges over all integers</u>). If $E_1 \oplus E_2 \oplus \ldots \oplus E_N \ne R_w$, then there is a vector x in R^∞ with $<g^n e_i,x> = 0$ for $1 \le i \le N$, and all n, and the conclusion follows.

Otherwise proceed as follows. For at least one i, $\dim E_i = \infty$. By reordering the e_i and E_i for $1 \le i \le N$ if necessary, we may assume $\dim E_i$ is finite for $1 \le i < M$ (perhaps $M = 1$), and $\dim E_i$ is infinite for $M \le i \le N$. Let $B = (v_1,v_2,\ldots,v_a)$ be an ordered basis for $E_1 \oplus \ldots \oplus E_{M-1}$.

Lemma 3.2. The set $B \cup \{g^n E_M\}$ is independent.

Proof. Otherwise, there is (at least one) relation

$$\sum_{j=\alpha}^{\beta} c_j g^j e_M \in \text{span } B, \qquad \alpha,\beta \in Z. \tag{3.3}$$

We may assume that c_α and c_β are non-zero. It is easy to show that $g^k e_M \in \text{span}(B \cup \{g^j e_M \mid \alpha < j < \beta\})$, first for $k = \beta+1, \beta+2, \ldots$, and then for $k = \alpha-1, \alpha-2, \ldots$. Thus E_M is contained in a finite dimensional set, a contradiction.

We return to the proof of the proposition.

3.4. Inductive assumption K. For $k = M, M+1, \ldots, K$, there are

(i) sets $B_k = \{g^j e_k\}$ of one of three types: empty, finite, in which case j ranges over an interval $0 \le j \le j_k$, or infinite, in which case j ranges over all integers, and

(ii) in case B_k is empty or finite, relations of the form

$$p_k(g) \in E_1 \oplus E_2 \oplus \ldots \oplus E_{k-1},$$

where $p_k(g)$ is a polynomial in g of degree $j_k + 1$, whose constant and j_k+1st degree terms are non-zero. Also,

(iii) $B \cup B_M \cup B_{M+1} \cup \ldots \cup B_K$ is independent, and

(iv) also a basis for $E_1 \oplus E_2 \oplus \ldots \oplus E_K$.

Lemma 3.2 yields the inductive assumption 3.4 for $K = M$. Assume now that 3.4 holds for some $K \ge M$. We shall verify 3.4 for $K+1$.

Either the set

$$B \cup B_M \cup \ldots \cup B_K \cup \{g^j e_{K+1} \mid 0 \le j < \infty\} \tag{3.5}$$

is independent, or there is a maximal independent subset of the form

$$B \cup B_M \cup \ldots \cup B_K \cup \{g^j e_{K+1} \mid 0 \le j \le j_{K+1}\}, \quad j_{K+1} = -1,0,1,\ldots, \tag{3.6}$$

and a polynomial $p_{K+1}(g)$ of degree exactly $j_{K+1}+1$ for which

$$p_{K+1}(g)e_{K+1} \in E_1 \oplus E_2 \oplus \dots \oplus E_K.$$

If the set (3.5) is independent, let $B_{K+1} = \{g^n e_{K+1}\}$. As usual, n ranges over Z. If $B \cup B_M \cup \dots \cup B_{K+1}$ were dependent, then for some finite power series in g,

$$f(g)e_{K+1} \in E_1 \oplus E_2 \oplus \dots \oplus E_K.$$

Then for some $a > 0$, $g^a f(g)$ is a polynomial in g; this yields a relation which contradicts the independence of (3.5). Thus the inductive assumption holds for $K+1$ if (3.5) is independent.

Suppose that (3.5) is dependent. Choose a maximal independent set (3.6); this yields B_{K+1} and p_{K+1}. If the constant term in p_{K+1} were non-zero, $g^{-1}p_{K+1}(g)$ would yield a relation contradicting the independence of the chosen set of the form (3.6). This yields parts (i), (ii) and (iii) of the inductive assumption 3.4 for $K+1$. Verification of part (iv) is similar to the proof of Lemma 3.2, and is omitted.

Continue until $K = N$.

3.7. <u>Construction of a useful basis for R^∞</u>. Let $B' = B \cup B_M \cup \dots \cup B_N$. Define an order $\prec$ on B' as follows:

$v_i \prec v_{i+1}$	for all i,
$v_i \prec g^n e_k$	for all i, all $k \geq M$, and all n,
$g^n e_k \prec g^n e_{k+1}$	for all $k \geq M$ and all n, and
$g^m e_k \prec g^n e_k$	for all $k \geq M$ whenever m preceeds n in the list $(0,1,-1,2,-2,3,\dots)$.

By construction, each R^n is contained in the span of some finite initial segment of B'. Define an orthonormal basis B'' for R_w by applying the Gramm-Schmidt process to B'. For each basis vector b' in B, let $\varphi(b')$ denote the corresponding basis vector in B''. Then

$$\langle b', \varphi(b') \rangle \neq 0, \quad \text{and} \quad \langle b', \varphi(b_1') \rangle = 0 \quad \text{if} \quad b' \prec b_1'.$$

It is easy to see that B'' is a basis for R^∞ as a topological vector space: by construction, for any e_i, $\langle e_i, b'' \rangle = 0$ for almost all b'' in B'', thus for any x in R^∞, the sum

$$\sum_{b'' \in B''} \langle b'', x \rangle b''$$

converges. It is easy to check that this sum converges to x, and that B'' is independent; thus B'' is a basis for R^∞.

3.8. <u>Construction of x.</u> There are three cases.

Case I: <u>For all k with $M \leq k \leq N$, B_k is infinite</u>. We can inductively define the coefficients in the sum

$$x = \sum_{b'' \in B''} c_{b''} b''$$

so that

$$\langle e_i, f^n x \rangle = \langle g^n e_i, x \rangle = \begin{cases} 0 & \text{if } 1 \leq i < M, \\ \varepsilon/3N & \text{if } M \leq i \leq N. \end{cases}$$

Proposition 3.1 follows.

Case II: <u>Although B_M is infinite, some other B_k with $M < k \leq N$ are finite, but for each polynomial p_k, the sum of coefficients ($p_k(1)$) is non-zero</u>. We shall define x and numbers δ_i, not all zero, so that

$$\langle e_i, f^n x \rangle = \langle g^n e_i, x \rangle = \begin{cases} 0 & \text{if } 1 \leq i < M, \\ \delta_i & \text{if } M \leq i < N. \end{cases}$$

This requires permuting an initial segment of B''. For example, if

$$p_k(g) e_k = \sum_{\substack{i<k \\ \text{finite}}} c_{n,i} g^n e_i, \tag{3.9}$$

we may define δ_k once all δ_i, $i < k$, and thus the right-hand-side of (3.9) have been defined. Choosing the independent δ_i non-zero, but so small that all $\delta_i < \varepsilon/2N$ yields Proposition 3.1.

Case III: <u>As in Case II, except that at least one</u> p_k <u>has coefficient-sum</u> $p_k(1) = 0$. Choose the largest $k = k_0$ for which this happens. We shall define x so that

$$\langle e_i, f^n x\rangle = \langle g^n e_i, x\rangle = \begin{cases} 0 & \text{if } 1 \le i < k_0, \\ \delta_{k_0} & \text{if } i = k_0, \\ \delta_i & \text{if } k_0 < i \le N. \end{cases} \tag{3.10}$$

As in Case II, we may have to permute an initial segment of B". The first line of (3.10) is easy to obtain. Because

$$\langle p_k(g)e_k, x\rangle = \sum_{\substack{i<k_0 \\ \text{finite}}} \langle c_{n,i} g^n e_i, x\rangle = 0$$

by construction, and $p_k(1) = 0$, we may choose δ_{k_0} to be an arbitrary non-zero number. By assumption on k_0, we may continue as in Case II. Again, the proposition follows, for suitable δ_{k_0} and δ_i.

This completes the proof of Proposition 3.1.

Our choice of N now implies that $|f^n x| < \varepsilon$ for all n (see Section 2). Because ε was arbitrary, f is not expansive.

<u>Remarks 3.11</u>. Let M be a 2×2 Anosov matrix. Then the infinite matrix

$$M' = \begin{pmatrix} M & I & & & & \\ & M & I & & & \\ & & M & I & & \\ & & & \cdot & & \\ & & & & \cdot & \\ & & & & & \cdot \end{pmatrix}$$

yields an expansive linear homeomorphism of $T^\infty_w = \cup T^n$. However, M' is not even an isomorphism of R^{∞}.

REFERENCES

1. D. Anosov, Geodesic flows on closed Riemann manifolds with negative curvature, Proc. Steklov Inst. Math., 90(1967).

2. J. Cohen, Inverse limits of principal fibrations, Proc. London Math. Soc., (3)27(1973), 178-192.

3. J. Franks, Anosov diffeomorphisms, Proc. Symp. Pure Math., 14 (1970), 61-93.

4. A. Manning, There are no new Anosov diffeomorphisms on tori, Amer. J. Math., 96(1974), 422-429.

HOFSTRA UNIVERSITY

SHAPE THEORY AND DYNAMICAL SYSTEMS

by

Harold M. Hastings*

We develop several relations between shape theory (Čhech homotopy theory), and the structure of attractors.

§1. Introduction

The definition of attractors, the wide use of Čech homology theory in dynamical systems (see, e.g., J. Alexander and J. Yorke [1], M. Shub [21], and R. Bowen [6], and many others in almost any issue of "Topology"), and W.C. Chewning's example of a continuous, non-differentiable flow [8], see Section 5, suggest an interesting interface between shape theory and dynamical systems. This paper is a preliminary report on that interface, preliminary in that more questions are raised than answered.

Basic shape theory is reviewed in Section 2, and compared with the Poincaré-Bendixson theorem in Section 3. Section 4 relates Steenrod homology and symbolic dynamics, and Section 5 raises some questions about smoothings.

We acknowledge helpful conversations with R. Bowen, C. Conley, H. Gluck and J. Yorke at the conference, also with D. Puppe and J. West.

§2. Shape Theory

We review the basic shape theory used in 3-5, below, see, e.g. K. Borsuk's book [5] and [10]. Shape theory and singular homology theory form complimentary approaches to the algebraic topology of compact metric spaces. Consider, for example, the <u>Warsaw circle</u> S_W

*Partially supported by National Science Foundation (MCS77-01628).

obtained by replacing an arc in a circle by the "sin 1/x" curve and its limit line.

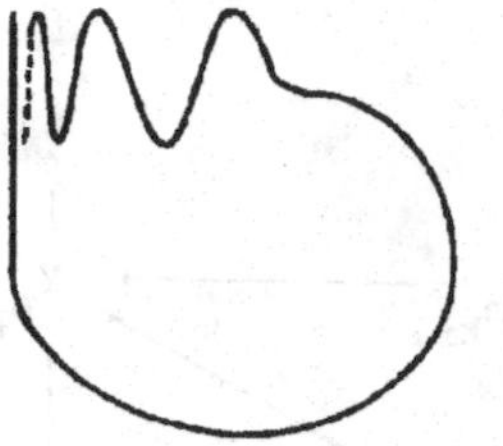

(2.1)

Although all singular invariants of S_W are trivial, there is an obvious essential map $S_W \to S^1$ which induces an isomorphism on Čech homology. In fact, S_W and S^1 have the same shape.

Two of the many definitions of shape seem particularly relevant.

(2.2) <u>Neighborhood systems</u>. Consider a compactum X embedded in a finite-dimensional manifold M with neighborhood basis

$$\{X_n\} = \{X_0 \supset X_1 \supset X_2 \supset \dots\}. \tag{2.3}$$

We require that each X_n be nice (e.g., a manifold). Then the tower $\{X_n\}$ (2.3) is the <u>shape</u> of X. This may be generalized to infinite-dimensional M or non-compact X, see [5].

(2.4) <u>Inverse limits</u>. As an obvious generalization of (2.3), assume that X is the inverse limit of a tower of polyhedra (or more generally ANR's)

$$\{X_n\} = \{X_0 \leftarrow X_1 \leftarrow X_2 \leftarrow \dots\}. \tag{2.5}$$

Again, $\{X_n\}$ is the shape of X. See S. Mardešić and J. Segal [17]. In particular, the shape of a polyhedron X is the constant tower with all $X_n = X$.

(2.6) <u>Shape equivalence</u>. Any two neighborhood bases $\{X_m\}$ and $\{X'_n\}$ of a compactum X (in a given manifold) are cofinal, that is, for each X'_n there is an $X_m \subset X'_n$ and vice-versa. More generally, compacta X and Y are said to have the same shape if there is a homotopy commutative diagram

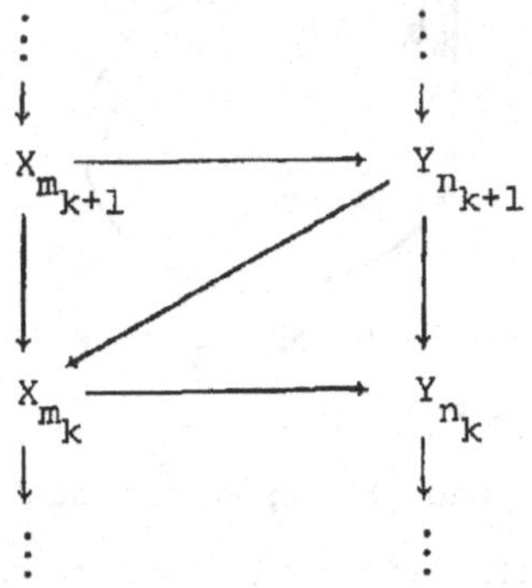

with $m_k \to \infty$, and $n_k \to \infty$ with k. This definition may be formalized by introducing a suitable category of inverse systems -- see M. Artin and B. Mazur [2], and D.A. Edwards, and the author [11].

(2.7) Pro-groups. The usual group-valued functors of algebraic topology yield functors into pro-groups, the category of inverse systems of groups, in shape theory. For example, the pro-homology of a compactum X is given by

$$\text{pro-}H_m(X) \quad = \quad \{H_m(X_n)\}_n .$$

Čech homology is obtained by applying the inverse limit lim to pro-homology. The more powerful Steenrod homology [23] ${}^{S}H_*$ requires either strong shape theory [11] or an equivalent geometric formulation [18]; there is a short-exact-sequence [23], [18], see [19] for $\lim^1$,

$$0 \longrightarrow \lim_n^1 \{H_{n+1}(X_n)\} \longrightarrow {}^{S}H_m(X) \longrightarrow \check{H}_m(X) \longrightarrow 0.$$

(2.8) Solenoids. The k-adic solenoid $S_{(K)}$ is the inverse limit of the tower

$$\{S^1 \xleftarrow{k} S^1 \xleftarrow{k} S^1 \xleftarrow{k} \cdots\}$$

of circles bonded by degree k (e.g., k^{th} power) maps. D.E. Christie [9] essentially computed

$$\check{H}_m(S_{(2)}) = Z \text{ if } m = 0, \text{ and } = 0, \text{ otherwise},$$

$$S_{H_0}(S_{(2)}) = Z_{(2)}, \text{ the 2-adic integers.}$$

(2.9) *Stable shape, movability*. A compactum X is said to have stable shape if $\{X_n\}$ is equivalent to a constant tower; X is called movable if for each n there is an $r > n$ such that for all $s > r$, the dotted arrow in

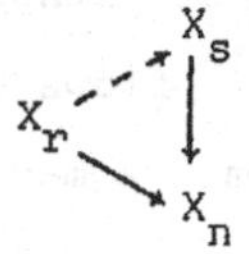

exists. For example, the k-adic solenoids $(k > 1)$ are neither stable nor movable, the infinite torus $T^\infty = \prod_{i=1}^{\infty} S^1$ is movable but not stable, and the Warsaw circle S_W is stable.

Pathological spaces, such as R. Bing's dog-bone space [3, 4], will be discussed briefly in Section 5.

§3. The Poincaré-Bendixson Theorem

We briefly relate the Poincaré-Bendixson theorem, more precisely the Corollary 3.1 below, and stability in shape theory. Details appear in [13].

(3.1) Corollary to the Poincaré-Bendixson theorem (e.g, [14, pp. 239-254]). *Given a C^1 flow $f = \{f_t\}$ on R^2 and an annulus $M = S^1 \times I$ in R^2 such that orbits through the boundary of M enter M as time t increases, and there are no rest points in M, then there is a limit cycle in the interior of M.*

A shape-theorist might begin a heuristic "proof" of Corollary

3.1 as follows. Let $M_t = f_t(M)$. Then the resulting tower $\{M_n\}$ consists of annuli and degree-one inclusions. Hence their intersection, the attractor $K = \bigcap_{n=0}^{\infty} M_n$ has the shape of a circle. An easy generalization yields the following.

Theorem 3.2 [13]. Let M be a compact, n-dimensional submanifold of R^n. Let $f = f_t$ be a continuous semiflow on M such that orbits through the boundary of M enter M for positive t. Then there is an attractor K in the interior of M, with K shape equivalent to M via the inclusion.

Remark 3.3 [13]. K itself can be pathological, e.g. $S_W \times S^1$ (S_W is the Warsaw circle), even for C^1 flows without rest points.

On the other hand, Theorem 3.2 may fail completely for diffeomorphisms $M_0 \to M_1 \subset \operatorname{int} M_0$; for example, the dyadic solenoid is obtained from a degree-2 diffeomorphism from the solid torus into itself.

§4. Solenoids.[1]

There is a close relation between the shape theory and the dynamical systems theory [24] of solenoids. In particular, some of our results appear closely related to shift equivalence versus strong shift equivalence, see [24].

(4.1) Generalized solenoids. There are three generalizations. Williams [24, p. 150] replaces S^1 by a finite wedge $\vee S^1$, and introduces matrix maps

$$\alpha\colon \bigvee_{i=1}^{n} S^1 \to \bigvee_{i=1}^{m} S^1, \qquad \alpha_* = A \text{ on } H_1,$$

[1] Some of this material overlaps results of R.F. Williams, Classification of one dimensional attractors, Global Analysis, Proc. Symp. in Pure Math. XIV, American Mathematical Society, 1970, Vol. 1, pp. 341-361.

where A is an $m \times n$ matrix over Z^+. If α realizes A, and β realizes B, then $\alpha\beta$ realizes AB provided that the composites are defined. Unfortunately, $RA = BR$ need not imply $\rho\alpha \simeq \beta\rho$.

J. Keesling [15] and W. Lawton [16] use a finite product $\prod S^1$ in place of William's wedge. Now an $m \times n$ matrix determines a unique matrix map as above, and $RA = BR$ implies $\rho\alpha \simeq \beta\rho$. Unfortunately the dimension has been raised (the Williams examples embed as dynamical systems in S^3), and symbolic dynamics on the m-torus, $m > 2$, T^m (see, e.g., Ya. Sinai [22]) is quite complex.

It appears interesting to replace S^1 by, say, S^3 in both definitions.

We now relate pro-invariants, Steenrod invariants, and Bowen-Franks [7] invariants.

(4.2) Bernoulli shifts. The full (Bernoulli) k-shift is realized by the k-adic solenoid 2.8. Thus distinct k-shifts can be distinguished by pro-homology

$$\text{pro-}H_1(S_{(k)}) = \{Z \xleftarrow{k} Z \xleftarrow{k} Z \xleftarrow{k} \cdots\}$$

(where $\xleftarrow{k}$ denotes multiplication by k) or Steenrod homology

$${}^{S}H_0(S_{(k)}) = \lim{}^1 \{Z \xleftarrow{k} Z \xleftarrow{k} Z \xleftarrow{k} \cdots\}.$$

(4.3) The functors lim and $\lim^1$. For an abelian group G, let (G,α) denote the pro-group

$$\{G \xleftarrow{\alpha} G \xleftarrow{\alpha} G \xleftarrow{\alpha} \cdots\}.$$

Then $\lim(G,\alpha)$ and $\lim^1(G,\alpha)$ are the kernel and cokernel, respectively, of the map

$$\delta\colon \prod_{i=1}^{\infty} G \to \prod_{i=1}^{\infty} G \quad \text{with} \quad \delta(g_0,g_1,g_2,\ldots) = (g_0-g_1,\ g_1-g_2,\ g_2-g_3,\ \ldots).$$

Thus the Bowen-Franks invariants $\ker(I-\alpha)$ and $\operatorname{coker}(I-\alpha)$ are essentially restrictions of $\lim(G,\alpha)$ and $\lim^1(G,\alpha)$.

We now consider pro-groups and shift equivalence.

(4.4) The strong shift equivalence $ST \sim TS$. In this case, the pro-groups (Z,ST) and (Z,TS) are cofinal in the pro-group

$$\{Z \xleftarrow{S} Z \xleftarrow{T} Z \xleftarrow{S} Z \xleftarrow{T} \ldots\},$$

roughly the "odd" and "even" terms, and hence, pro-isomorphic. Williams [24] realizes this isomorphism as an isomorphism of dynamical systems on S^3.

(4.5) Shift equivalence. Williams [24] calls matrices A and B shift equivalent if there are matrices S and T over Z^+, and a positive integer k, with

$$SA = BS, \quad AT = TB, \quad TS = A^k, \quad \text{and} \quad ST = B^k.$$

Theorem 4.6 (compare [7, Prop. (1.1)]). *If A and B are shift equivalent, then the pro-groups (Z^m,A) and (Z^n,B) are pro-isomorphic, compatibly with the shifts on (Z^m,A) and (Z^n,B).*

Proof. This is easier than [7, Prop. (1.1)]. Consider the diagram

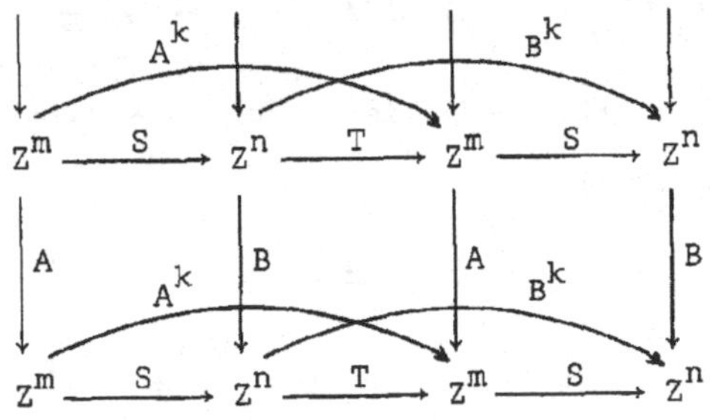

The required pro-isomorphism is given by

$$\sigma: (Z^m,A) \to (Z^n,B), \qquad sh^k \circ \tau: (Z^n,B) \to (Z^m,A),$$

where σ is defined by applying S levelwise, τ by applying T levelwise, and sh is the shift (and pro-isomorphism)

$$sh(g_0,g_1,g_2,\ldots) = (g_1,g_2,g_3,\ldots).$$

Essentially, $sh = A^{-1}$. The conclusion follows easily.

It is easy to prove the following.

Proposition 4.7. _Let_ A _and_ B _be shift equivalent, where_ A _has no eigenvalues of modulus_ 0 _or_ 1. _Then_ A _and_ B _induce equivalent Anosov diffeomorphisms._

Question 4.8. Can you recover the symbolic dynamics of A and B from (4.7)? Compare [22].

§5. Smoothings

W.C. Chewning [8] constructed a continuous flow not conjugate to a C^1 flow; J. Harrison [12] found C^r flows not conjugate to C^{r+1} flows for all $r > 0$ by different techniques. We shall briefly discuss the question: can such flows by smoothed by forming suitable products? More precisely, if f_t is such an example, and g_t is say the constant flow on R, or a flow with 0 a global attractor on R, is $f_t \times g_t$ conjugate to a smoother flow?

Chewning lets f_t be the translation flow on $B \times R$, where B is R.H. Bing's [3,4] dog-bone space: $B \times R \cong R^4$, but B is _not_ a manifold. Then f_t has non-manifold cross-sections, and hence is not conjugate to a C^1 flow. We may replace B by its one-point compactification B^*, and R by S^1 to obtain a compact "Bing-Chewning" example. It is easy to prove the following.

Theorem 5.1. _Let_ f_t _be the flow on_ $B^* \times S^1$ _obtained from the translation on_ S^1, _let_ g_t _be the constant flow on_ R, _and let_ h_t _be a smooth flow on_ R _with unique global attractor_ 0. _Finally, let_ f'_t

be the flow on $S^3 \times S^1$ *obtained from translation on* S^1. *Then* $f_t \times g_t$ *and* $f'_t \times g_t$ *are topologically conjugate, and* $f_t \times h_t$ *and* $f'_t \times h_t$ *are locally but not globally conjugate.*

For the last statement, consider the restrictions of the flows to the attractors $B \times S^1$ and $S^3 \times S^1$, respectively. It appears likely that $f_t \times h_t$ is not conjugate to a C^1-flow. In fact, a related flow, the suspension of the involution $id \times \tau$, $\tau x = -x$, on $B \times R^2$, cannot be smoothed to a C^2 flow by forming a product with h_t. We conclude with a probably difficult conjecture: Harrison's examples cannot be embedded in smoother flows by forming products with h_t.

REFERENCES

1. J. Alexander and J. Yorke, Global bifurcation of periodic orbits, (to appear in *Amer. J. Math.*).
2. M. Artin and B. Mazur, *Étale homotopy*, Lecture Notes in Math. 100, Springer-Verlag, Berlin-Heidelberg-New York, 1969.
3. R.H. Bing, A decomposition of E^3 into points and tame arcs such that the decomposition space is topologically different from E^3, *Ann. of Math.* (2) 65(1957), 484-500.
4. R.H. Bing, The Cartesian product of a certain nonmanifold and a line is E^4, *Ann. of Math.* (2) 70(1959), 399-412.
5. K. Borsuk, *Shape Theory*, Monographie Matematyczne, Warszawa, 1976.
6. R. Bowen, On Axiom A diffeomorphisms, (to appear in CBMS/NSF Conference Proceedings).
7. R. Bowen, and J. Franks, Homology for zero-dimensional nonwandering sets, (to appear).
8. W.C. Chewning, A dynamical system on E^4 neither isomorphic nor equivalent to a differentiable system, *Bull. Amer. Math. Soc.* 80 (1974), 150-153.
9. D. Christie, Net homotopy theory for compacta, *Trans. Amer. Math. Soc.* 56(1944), 275-308.
10. D.A. Edwards, Étale homotopy and shape, in Algebraic and geometric methods in topology, ed., L. McAuley, Lecture Notes in Math. 428, Springer-Verlag, Berlin-Heidelberg-New York, 1974, 58-107.

11. D.A. Edwards, and H.M. Hastings, Čech and Steenrod Homotopy Theory, With Applications to Geometric Topology, Lecture Notes in Math. 542, Springer-Verlag, Berlin-Heidelberg-New York, 1976.

12. J. Harrison, Unstable diffeomorphisms, Annals of Math. (2) 102 (1975), 85-94.

13. H.M. Hastings, A higher dimensional Poincaré-Bendixson theorem, (preprint).

14. M.W. Hirsch and S. Smale, Differential Equations, Dynamical Systems, and Linear Algebra, Academic, New York, 1974.

15. J. Kneesling, The Čech homology of compact connected abelian topological groups with applications to shape theory, in Geometric Topology, Proceedings 1974, ed. L.C. Glaser and T.B. Rushing, Lecture Notes in Math. 438, Springer-Verlag, Berlin-Heidelberg-New York, 1975.

16. W. Lawton, The structure of compact connected groups which admit an expansive automorphism, in Recent Advances in Topological Dynamics, ed. A. Beck, Lecture Notes in Math. 318, Springer-Verlag, Berlin-Heidelberg-New York, 1973, 182-196.

17. S. Mardešić and J. Segal, Shapes of compacta and ANR systems, Fund. Math. 72(1971), 41-59.

18. J. Milnor, On the Steenrod homology theory, (mimeo. notes), Berkeley, 1961.

19. J. Milnor, On axiomatic homology theory, Pacific J. of Math. 12 (1962), 337-341.

20. W. Parry and D. Sullivan, A topological invariant of flows on one-dimensional spaces, Topology 14(1975), 297-299.

21. M. Shub, Dynamical systems, filtrations, and entropy, Bull. Amer. Math. Soc. 80(1974), 27-41.

22. Ya. G. Sinai, Introduction to Ergodic Theory, Princeton University Press, 1977.

23. N.E. Steenrod, Regular cycles on compact metric spaces, Ann. of Math. (2) 41(1940), 833-851.

24. R.F. Williams, Classification of subshifts of finite type, Ann. of Math. (2) 98(1973), 120-153, Errata, ibid., 99(1974), 380-391.

HOFSTRA UNIVERSITY

ON A THEOREM OF SELL

by

Russell A. Johnson

Introduction

This paper outlines an alternate approach to a theorem of Sell ([13]) concerning the existence and structure of stable, unstable, and branch manifolds for certain non-linear ordinary differential equations $\dot{x} = A(t)x + F(t,x)$ (see 1.2 for assumptions on A and F). Details will appear elsewhere; no proofs are given here. We use theorems on pseudo-hyperbolic linear maps contained in papers by Hirsch-Pugh-Shub ([7]) and Irwin ([8]). Results of Foster ([3]) are also of importance.

It will be seen that we need assumptions which are a bit stronger than Sell's (Section 2). Specifically, our non-linear perturbations $F(t,x)$, together with all derivatives $D_2{}^jF$ of interest, must be uniformly continuous in t; Sell does not require this. On the other hand, our "invariant manifolds" are actual topological manifolds in the usual sense, and are C^1 if F is. If the invariant manifold is a "λ_i-local stable manifold" (see (v) in the discussion beginning in Section 2) with $\lambda_i \leq 0$, then it is C^r if F is C^r. Moreover, orbits originating in a λ_i-local stable manifold are defined for all $t \geq 0$, and are contracted to zero exponentially as $t \to \infty$. Similar statements hold for λ_i-local unstable manifolds if $\lambda_i \geq 0$.

§1. Preliminaries

Notation 1.1. Let $X = \mathbb{R}^n$ (we think of an ODE on $\mathbb{C}^n$ as an ODE on $\mathbb{R}^{2n}$). Give X the norm $|\cdot|$ induced by the usual inner product $< , >$. Let $L(X,X)$ be the set of linear operators on X; let $L^r(X,X)$ be the space of r linear maps from X to itself with norm

$$|S| = \sup_{\substack{|x_i|<1 \\ 1\leq i\leq r}} |S(x_1,\ldots,x_r)|.$$

Assumptions, Notation 1.2. We will consider equations

(1) $\dot{x} = A(t)x \quad (t \in \mathbb{R}, \quad x \in X)$,

and

(2) $\dot{x} = A(t)x + F(t,x)$.

Here $A: \mathbb{R} \to L(X,X)$ is uniformly bounded and uniformly continuous. The map $F: R \times X \to X$ (a) is C^r in x $(r \geq 0)$; (b) satisfies $F(t,0) = 0$ for all t. It also has the following properties. (c) If $K \subset X$ is compact, then $F|_{\mathbb{R}\times K}$ is uniformly bounded (u.b.) and uniformly continuous (u.c.) (d) If $r > 0$, then all x-derivatives $D_2^j F|_{\mathbb{R}\times K}$ are u.b. and u.c. $(1 \leq j \leq r)$. (e) If $\varepsilon > 0$, $\exists\, \delta > 0$ such that $|x|,|y| < \delta$ implies $|F(t,x) - F(t,y)| < \varepsilon|x-y|$ $\forall\, t$.

We will also consider equations

$(2)_p$ $\dot{x} = A(t)x + \varphi_p(x)F(t,x)$,

where $p > 0$ and $\varphi_p: X \to X$ is a C^∞ function equal to 1 if $|x| \leq p$ and equal to 0 if $|x| \geq 2p$.

Proposition 1.3. There exists a flow $(\Omega,\mathbb{R})$, with Ω compact metric, and continuous functions $a: \Omega \to L(X,X)$ and $f: \Omega \times X \to X$ with the following properties.

(a) There exists $\omega_0 \in \Omega$ such that $\{\omega_0 \cdot t \mid t \in \mathbb{R}\}$ is dense in Ω, and $a(\omega_0 \cdot t) = A(t)$, $f(\omega_0 \cdot t, x) = F(t,x)$ $(t \in \mathbb{R}, \ x \in X)$. (If the flow is given by the map $\Phi: \Omega \times \mathbb{R} \to \Omega$, and let $\omega_0 \cdot t \equiv \Phi(\omega_0, t)$.)

(b) For all $\omega \in \Omega$, $f(\omega,0) = 0$.

(c) Given $\varepsilon > 0$, there exists $\delta > 0$ such that, if $|x|,|y| < \delta$, then $|f(\omega,x) - f(\omega,y)| < \varepsilon|x-y|$ $(\omega \in \Omega)$.

(d) <u>If</u> F <u>is</u> C^r <u>in</u> x $(r \geq 1)$, <u>then</u> $f(\omega,x)$ <u>is</u> C^r <u>in</u> x. <u>If</u> $K \subset X$ <u>is compact, then</u> $D_2^j f\colon \Omega \times K \to L^j(X,X)$ <u>is continuous</u> $(1 \leq j \leq r)$. (Here "D_2" means "derivative with respect to the second variable.")

<u>Remark 1.4</u>. We see that equation (1) is contained in the class of equations

$$(1)_\omega \quad \dot{x} = a(\omega\cdot t)x \qquad (\omega \in \Omega);$$

in fact, $(1) = (1)_{\omega_0}$. Similarly, $(2) = (2)_{\omega_0}$, where

$$(2)_\omega \quad \dot{x} = a(\omega\cdot t)x + f(\omega\cdot t,x) \qquad (\omega \in \Omega);$$

and $(2)_p = (2)_{p,\omega_0}$, where

$$(2)_{p,\omega} \quad \dot{x} = a(\omega\cdot t)x + \varphi_p(x)f(\omega\cdot t,x) \qquad (\omega \in \Omega).$$

<u>Assumptions 1.5</u>. (a) Each equation $(2)_\omega$ has a unique noncontinuable solution for each initial condition $x(0) = x_0$.

(b) Let $\omega_n \to \omega$, $x_n \to x_0$. If $\varphi_n(t)$ $(\varphi(t))$ is the unique noncontinuable solution to $(2)_{\omega_n}$ $((2)_\omega)$ such that $\varphi_n(t_0) = x_n$ $(\varphi(t) = x_0)$, then $\varphi_n(t) \to \varphi(t)$ uniformly for t in compact subsets of the interval of definition of φ.

Assumptions 1.5a) and b) ensure that the objects $\tilde{N}$ and N defined in 1.6 below really are flows. See ([11], p. 14).

<u>Definitions 1.6</u>. Consider now the trivial vector bundle $V = \Omega \times X$. Let $\pi\colon (\omega,x) \to \omega$ be the projection. Give V the trivial Finsler: $|(\omega,x)| = |x|$.

Define a flow L on V as follows: $(\omega,x_0)\cdot t = (\omega\cdot t,x(t))$, where $x(t)$ is the solution to $(1)_\omega$ with initial condition $x(0) = x_0$. We think of L as "generated" by equation (1). Observe that $(\omega, \alpha x_1+\beta x_2)\cdot t = (\omega\cdot t, \alpha x_1(t)+\beta x_2(t))$. Thus L is linear on fibers; it is a <u>linear skew-product flow</u> ([11], [12]). We will write L_t for the map $(\omega,x_0) \to (\omega,x_0)\cdot t \quad (t \in \mathbb{R})$.

Similarly define a local flow $\tilde{N}$ (for Nonlinear) on $\Omega \times X$ via $(\omega,x_0)\cdot t = (\omega\cdot t, x(t))$, where $x(t)$ solves $(2)_\omega$ with $x(0) = x_0$. This flow is local since solutions may not be defined for all time. Finally, if p is fixed, define a flow N by $(\omega,x_0)\cdot t = (\omega_0\cdot t, x(t))$, where $x(t)$ solves $(2)_{\omega,p}$ with $x(0) = x_0$. This is a flow; solutions to $(2)_{\omega,p}$ exist for all time. (The flow N depends on p; for convenience, we write N instead of N^p). Let $\tilde{N}_t$, N_t be the time-t maps. Observe that $\tilde{N}_t(v)$ is defined and equals $N_t(v)$ on $(-t_0,t_0)$ whenever $|\tilde{N}_t(v)| \le p$ on $(-t_0,t_0)$ $(v \in V)$.

Next we consider the spectrum of L ([11], [12]).

Definitions 1.7. Let $\lambda \in \mathbb{R}$. Say that λ is in the resolvent of L if there are constants K, $\alpha > 0$ and continuous subbundles ([12]) $V_s, V_u \subset V$ such that

(a) $V = V_s \oplus V_u$ (Whitney sum);

(b) $L_t(V_s) \subset V_s$, $L_t(V_u) \subset V_u \quad (t \in \mathbb{R})$;

(c) if $v \in V_s$, then $e^{-\lambda t}|L_t v| \le Ke^{-\alpha t}|v| \quad (t \ge 0)$;

(d) if $v \in V_u$, then $e^{-\lambda t}|L_t v| \le Ke^{\alpha t}|v| \quad (t \le 0)$.

If λ is not in the resolvent of L, it is in the spectrum of L, $Sp(L)$.

Spectral Theorem 1.8 ([12], [13]). The spectrum of L may be written as

$$\bigcup_{i=1}^{n+1} [a_i,b_i] \quad (n \ge 0), \quad \text{where} \quad -\infty < a_1 \le b_1 < a_2 \le \dots < a_{n+1} \le b_{n+1} < \infty.$$

We may write $V = V_1 \oplus V_2 \oplus \dots \oplus V_{n+1}$ (Whitney sum), where the V_i are continuous subbundles of V, invariant under each L_t. Moreover, $Sp(L|_{V_i}) = [a_i,b_i]$.

Definitions 1.9. From now until 2.22(b), fix numbers $\lambda_1,\dots,\lambda_n$ such that $b_i < \lambda_i < a_{i+1}$ $(1 \le i \le n)$.

§2.

We first outline the developments in Section 2. Consider the linear skew-product flow L induced by equation (1). Let $[a_1,b_1],\dots,[a_{n+1},b_{n+1}]$ be the spectral intervals (1.8), and choose $\lambda_i \in (a_i,b_{i+1})$ $(1 \le i \le n)$. Let $S_i{}^L(\omega) = \pi^{-1}(\omega) \cap [V_1 \oplus \dots \oplus V_i]$.

$$U_i{}^L(\omega) = \pi^{-1}(\omega) \cap [V_{i+1} \oplus \dots \oplus V_{n+1}],$$

$$Q_i{}^L(\omega) = \pi^{-1}(\omega) \cap V_i = S_i{}^L(\omega) \cap U^L_{i-1}(\omega).$$

These are all vector spaces; $S_i{}^L(\omega)$ $(U_i{}^L(\omega))$ is the set of vectors in $\pi^{-1}(\omega)$ which "grow no faster than $e^{\lambda_i t}$" as $t \to \infty$ (as $t \to -\infty$). And $Q_i{}^L(\omega)$ is the set of vectors which grow no faster than $e^{\lambda_i t}$ $(e^{\lambda_{i-1} t})$ as $t \to \infty$ (as $t \to -\infty$).

We seek local analogues $S_i(\omega,\delta)$, $U_i(\omega,\delta)$, $Q_i(\omega,\delta)$ of these spaces for nonlinear equations $(2)_\omega$. They will be found as follows (we consider the S_i's first).

(i) For fixed λ_i, consider equations $(2)_{\omega,p}$ for small p. Let $E = \{\sigma\colon \Omega \to V \mid \sigma$ is a bounded section$\}$. Then L_t, N_t induce mappings ℓ_t, n_t of E into E $(t \in R)$. Each ℓ_t is $e^{\lambda_i t}$ -- hyperbolic (2.1), with splitting $E = E_1 \oplus E_2$, where

$$E_1 = \{\sigma \mid \sigma(\Omega) \subset V_1 \oplus \dots \oplus V_i \equiv V_s\},$$

$$E_2 = \{\sigma \mid \sigma(\Omega) \subset V_{i+1} \oplus \dots \oplus V_{n+1} \equiv V_u\}.$$

(ii) For small p, results of ([7]), ([8]) imply that there is a function $h_1\colon E_1 \to E_2$ such that graph

$$(h_1) = \{\sigma \in E \mid e^{-\lambda_i t}\|N_t\sigma\| \to 0 \text{ as } t \to \infty\}.$$

Also, h_1 is Lipschitz: if $F(t,x)$ is C^r in x, then h_1 is C^1, and is C^r if $\lambda_i \leq 0$.

(iii) One then shows that $h_1(\sigma)(\omega) = H_1(\sigma(\omega))$, where $H_1\colon V_s \to V_u$ is a fiber-preserving map (i.e., $\pi \circ H_1 = \pi$). by ([3]), each map $H_{1,\omega} \equiv H_1|_{V_s \cap \pi^{-1}(\omega)}$ is Lipschitz, C^1, or C^r if h_1 is.

(iv) Let $S_i(\omega) = \text{graph}\,(H_{1,\omega})$ $(\omega \in \Omega)$. One sees that

$$S_i(\omega) = \{v \in \pi^{-1}(\omega) \mid e^{-\lambda_i t}|N_t v| \to 0 \text{ as } t \to \infty\}.$$

Also, $S_i(\omega)$ is a topological manifold; it is C^1 if $F(t,x)$ is C^r in x, and is C^r if $\lambda_i \leq 0$.

(v) Finally, the "λ_i-local stable manifold" $S_i(\omega,\delta)$ of $(2)_\omega$ is obtained by intersecting $S_i(\omega)$ with a small neighborhood of the zero section in V.

One obtains $U_i(\omega)$, $U_i(\omega,\delta)$ in a similar way (these will be C^r if $\lambda_i \geq 0$). Finally, one defines $Q_i(\omega) = S_i(\omega) \cap U_{i-1}(\omega)$ and $Q_i(\omega,\delta) = S_i(\omega,\delta) \cap U_{i-1}(\omega,\delta)$.

Definition 2.1. Let E be a Banach space, $T\colon E \to E$ a linear endomorphism. Suppose $E = E_1 \oplus E_2$ (E_k closed), where $T(E_k) \subset E_k$ $(k = 1,2)$. Let $T_1 \equiv T|E_1$, $T_2 \equiv T|E_2$, and suppose that, for some norm and some ρ, one has $\|T_1\| < \rho$, $\|T_2^{-1}\|^{-1} > \rho$. Then T is *ρ-hyperbolic*.

We state two results concerning ρ-hyperbolic maps, In proving these theorems, it is assumed that $\|x\| = \max(\|x_1\|, \|x_2\|)$. This is no real restriction on $\|\cdot\|$, since by the closed graph theorem the projections $E \to E_k$ are continuous.

Theorem 2.2 ([8], Theorems 29 and 30). *Let* $T\colon E = E_1 \oplus E \to E$ *be a ρ-hyperbolic automorphism. Let* $\varepsilon = \min(\rho - \|T_1\|, \|T_2^{-1}\|^{-1} - \rho)$. *Let* $f\colon E \to E$ *satisfy* $\text{Lip}(f-T) < \varepsilon$, $f(0) = 0$.

There is a unique Lipschitz map $h: E_1 \to E_2$ such that, for all $x \in E$, $x \in$ graph h if and only if $\rho^{-n}\|f^n(x)\|$ stays bounded as $n \to \infty$. "Stays bounded" may be replaced by "approaches zero."

Theorem 2.3 ([7], Theorem 5.1 and Corollary 5.3). Let T, E be as in 5.3; let $f: E \to E$ be C^r $(r \geq 1)$, with $f(0) = 0$. Let

$$W_1 = \{x \in E \mid \rho^{-n}\|f^n(x)\| \text{ stays bounded as } n \to \infty\},$$

$$W_2 = \{x \in E \mid \text{there exist inverse images } f^{-n}x \text{ such that } \rho^n\|f^{-n}x\| \text{ stays bounded as } n \to \infty\}.$$

There exists $\varepsilon > 0$ such that, if $\mathrm{Lip}(f-T) < \varepsilon$:

(a) there exist unique C^1 functions $h_1: E_1 \to E_2$ and $h_2: E_2 \to E_1$ such that $W_k = \text{graph } h_k$ $(k = 1,2)$;

(b) if $\|T_2\|^{-1}\|T_1\|^j < 1$ $(1 \leq j \leq r)$, then h_1 is C^r, and if $\|T_2{}^{-1}\|^j\|T_1\| < 1$ $(1 \leq j \leq r)$, then h_2 is C^r.

(c) if $Df(0) = T$, then W_k is tangent to E_k at 0.

We may replace "stays bounded" by "approaches zero."

Remarks 2.5. (a) In our applications, the function f in 2.3 will be invertible. Hence $W_2 = \{x \in E \mid \rho^n\|f^{-n}(x)\| \to 0 \text{ as } n \to \infty\}$.

(b) Observe that $\|T_2{}^{-1}\| < \frac{1}{\rho} < \|T_1\|^{-1}$. Assuming f invertible, we obtain from 2.3 that $\{x \in E \mid \rho^n\|f^{-n}(x)\| \to 0 \text{ as } n \to \infty\}$ is the graph of a Lipschitz function h_2 if $\mathrm{Lip}(f^{-1} - T^{-1})$ is small enough.

The Banach space E to which 2.3, 2.4 will be applied is the space of bounded sections $\sigma: \Omega \to V$.

2.5. From now until 2.18, we fix attention on some one λ_i. By 1.7, $V = V_s \oplus V_u$, where V_s (V_u) is contracted exponentially by $e^{-\lambda_i t}L_t$ as $t \to \infty$ (as $t \to -\infty$). Also, $V_s = V_1 \oplus \dots \oplus V_i$, $V_u = V_{i+1} \oplus \dots \oplus V_{n+1}$. We will often fail to indicate that quantities depend on i (we already have with V_s and V_u).

Definition 2.6. We put a new Finsler on V. Let $< , >$ be the inner product on X. Let $[a_1,b_1],\dots,[a_{n+1},b_{n+1}]$ be the spectral intervals (1.8). Choose $\beta_1 \in (-\infty,a_1)$, $\beta_2 \in (b_i,\lambda_i)$ (index i is fixed until 2.17). Assign an inner product to each fiber $\pi^{-1}(\omega) \cap V_s$ as follows:

$$(3)\quad \langle v,w\rangle_s = \int_0^\infty e^{2\beta_1 t}\langle L_{-t}v,L_{-t}w\rangle\,dt + \int_0^\infty e^{-2\beta_2 t}\langle L_t v,L_t w\rangle\,dt$$

$$(v,w \in V_s;\ \pi(v) = \pi(w)).$$

We also assign inner products to fibers $\pi^{-1}(\omega) \cap V_u$. Let $\beta_3 \in (\lambda_i,a_{i+1})$, $\beta_4 \in (b_{n+1},\infty)$. Define

$$(4)\quad \langle v,w\rangle_u = \int_0^\infty e^{2\beta_3 t}\langle L_{-t}v,L_{-t}w\rangle\,dt + \int_0^\infty e^{-2\beta_4 t}\langle L_t v,L_t w\rangle\,dt$$

$$(v,w \in V_u;\ \pi(v) = \pi(w)).$$

If $v \in V_s$ (V_u), let $|v|_s = \sqrt{\langle v,v\rangle_s}$ $(|v|_u = \sqrt{\langle v,v\rangle_u})$. For arbitrary $v \in V$, write $v = v_1 + v_2 \in V_s \oplus V_u$, and let $|v|_* = \max(|v_1|_s,|v_2|_u)$.

Definitions 2.7. Let E be the space of all bounded sections $\sigma\colon \Omega \to V$. Let $E_s = \{\sigma \in E \mid \sigma(\Omega) \subset V_s\}$, $E_u = \{\sigma \in E \mid \sigma(\Omega) \subset E_u\}$. Then $E = E_s \oplus E_u$; E_s and E_u are closed. If $\sigma \in E_s$, let $\|\sigma\| = \sup_{\omega\in\Omega}|\sigma(\omega)|_s$; if $\sigma \in E_u$, let $\|\sigma\| = \sup_{\omega\in\Omega}|\sigma(\omega)|_u$; if $\sigma = \sigma_1 + \sigma_2 \in E_s \oplus E_u$, let $\|\sigma\| = \max(\|\sigma_1\|,\|\sigma_2\|)$.

We could also norm E by $\|\sigma\|_1 = \sup_{\omega\in\Omega}|\sigma(\omega)|$, $\|\sigma\|_2 = \sup_{\omega\in\Omega}|\sigma|_*$.

Lemma 2.9. $\|\cdot\|$, $\|\cdot\|_1$, and $\|\cdot\|_2$ are equivalent.

Definitions 2.10. If $t \in \mathbb{R}$, define $\ell_t,\tilde{n}_t,n_t\colon E \to E$:

$$(\ell_t\sigma)(\omega) = L_t(\sigma(\omega\cdot -t)), \qquad (\tilde{n}_t\sigma)(\omega) = \tilde{N}_t(\sigma(\omega\cdot -t)),$$

$$(n_t\sigma)(\omega) = N_t(\sigma(\omega\cdot -t)).$$

In defining n_t, we assume p and φ_p are fixed: see 1.6.

The next lemma follows from 2.7.

Lemma 2.11. _For each_ $t \in \mathbb{R}$, ℓ_t _is_ $e^{\lambda_i t}$_-hyperbolic, with corresponding splitting_ $E_s \oplus E_u$.

Lemma 2.12. _Given_ $\varepsilon > 0$ _and_ $t_0 \in \mathbb{R}$, _there exists_ $\bar{p}$ _such that, if_ $p \leq \bar{p}$, _there is a_ C^∞ _bump function_ φ_p _(which generates a flow_ n_t _via equations_ $(2)_{\omega,p}$_) such that_ $\mathrm{Lip}(\ell_t - n_t) < \varepsilon$ $(|t| \leq t_0)$.

Lemma 2.13. _If_ $F(t,x)$ _is_ C^r _in_ x, _then_ n_t _is_ C^r _for any_ p _and_ φ_p.

2.14. Pick $t_0 > 0$, and consider the map ℓ_{t_0}. It is $e^{\lambda_i t_0}$-hyperbolic, and its inverse ℓ_{-t_0} is $e^{-\lambda_i t_0}$-hyperbolic (2.11). Suppose ε is so small that 2.2, 2.3, and 2.4(b) may be applied to $f\colon E \to E$ if $\mathrm{Lip}(f-T) < \varepsilon$ and $\mathrm{Lip}(f^{-1}-T^{-1}) < \varepsilon$. Using 2.12, choose p and φ_p such that the corresponding n_{t_0} satisfies $\mathrm{Lip}(\ell_{t_0} - n_{t_0}) < \varepsilon$, $\mathrm{Lip}(\ell_{-t_0} - n_{-t_0}) < \varepsilon$. Let $W_1 = \{x \in E \mid e^{-\lambda_i t} n_t(x)$ is bounded as $t \to \infty\}$, $W_2 = \{x \in E \mid e^{-\lambda_i t} n_t(x)$ is bounded as $t \to -\infty\}$. Both W_1 and W_2 are invariant under n_t for all t.

Proposition 2.15. (a) _There exist Lipschitz functions_ $h_1\colon E_1 \to E_2$ _and_ $h_2\colon E_2 \to E_1$ _such that_ $W_k = $ _graph_ h_k $(k = 1,2)$.

(b) _If_ $F(t,x)$ _is_ C^1 _in_ x, _so are the_ h_k, _and_ $Dh_k(0) = (0)$.

(c) _If_ $F(t,x)$ _is_ C^r _in_ x, _and if_ $\lambda_i \leq 0$, _then_ W_1 _is_ C^r; _if_ $\lambda_i \geq 0$, _then_ W_2 _is_ C^r.

The proof uses 2.2, 2.3, 2.13, and 2.14.

Remark 2.16. We may replace "is bounded" by "$\to 0$" in the definitions of W_1 and W_2.

Proposition 2.17. *Let* h_1, h_2 *be as in* 2.15.

(a) *There is a uniformly Lipschitz bundle map* $H_1: V_s \to V_u$ $(H_2: V_u \to V_s)$ *such that* $h_1(\sigma) = H_1 \circ \sigma$ $(h_2 = H_2 \circ \sigma)$.

(b) *If* h_k *is* C^r, *then* H_k *is* C^r *on each fiber* $\pi^{-1}(\omega)$.

The proof uses the technique of ([6], p. 150), together with results of Foster ([3]).

2.18. We now let i vary. First fix t_0 independent of i. Then, choose ε_i as in 2.15, $\overline{p}_i$ as in 2.13 (with $\varepsilon = \varepsilon_i$). Let $p_0 = \min_i \overline{p}_i$. Let N_t, n_t refer to the flows generated by equations $(2)_{\omega,p_0}$ (we can choose φ_{p_0} so that $\mathrm{Lip}(\ell_t - n_t) < \varepsilon_i$ for all i and all $|t| \le t_0$). Let $S_i(\omega) = \{v = (\omega,x) \in V \mid e^{-\lambda_i t} N_t(v) \to 0 \text{ as } t \to \infty\}$, $U_i(\omega) = \{v = (\omega,x) \in V \mid e^{-\lambda_i t} N_t(v) \to 0 \text{ as } t \to -\infty\}$. If $v \in S_i(\omega)$, then $N_t v \in S_i(\omega \cdot t)$; hence the family $\{S_i(\omega)\}$ is invariant. Similarly for $\{U_i(\omega)\}$. The $S_i(\omega)$ are the λ_i-*stable manifolds* for $(2)_{\omega,p_0}$, the $U_i(\omega)$ are the λ_i-*unstable manifolds*. By 2.13, 2.14, 2.15, and 2.16(a), one has $S_i(\omega) = \text{graph }(H_1 | V_S \cap \pi^{-1}(\omega))$, $U_i(\omega) = \text{graph }(H_2 | V_s \cap \pi^{-1}(\omega))$, where H_1 and H_2 depend on i.

2.19. We may define analogues of the bundles V_i as follows. Let $Q_i(\omega) = S_i(\omega) \cap U_{i-1}(\omega)$ $(1 \le i \le n+1;$ let $U_0(\omega) = X)$. The $Q_i(\omega)$ are the *branch manifolds* for equations $(2)_{\omega,p_0}$ ([13]). Note $Q_i(\omega) = \{(\omega,x) \in V \mid e^{-\lambda_i t} N_t v \to 0 \text{ as } t \to \infty \text{ and } e^{-\lambda_{i-1} t} N_t v \to 0 \text{ as } t \to -\infty\}$.

2.20. We now consider results for equations $(2)_\omega$. Choose p_0 and φ_{p_0} as in 2.21. In 2.7 we defined a Finsler $|\cdot|_*$ on V: in what follows, let $V(\delta) = \{v \in V \mid |v|_* < \delta\}$. Let $S_i(\omega,\delta) = S_i(\omega) \cap V(\delta)$;

similarly define $U_i(\omega,\delta)$, $Q_i(\omega,\delta)$. Recall (1.7) that $\tilde{N}_t$ refers to the local flow on V generated by solutions of equations $(2)_\omega$. The $S_i(\omega,\delta)$ are the λ_i-local stable manifolds for $(2)_\omega$; the $U_i(\omega,\delta)$ are the λ_i-local unstable manifolds.

Theorem 2.21. There exists a $\delta > 0$ with the following properties.

(a) $S_i(\omega,\delta)$, $U_i(\omega,\delta)$ are Lipschitz manifolds. If $F(t,x)$ is C^r in x, they are C^1. If $\lambda_i \leq 0$, $S_i(\omega,\delta)$ is C^r; if $\lambda_i \geq 0$, $U_i(\omega,\delta)$ is C^r.

(b) If $|\tilde{N}_t v|_* \leq \delta$ for all t in some interval I containing 0, and if $v \in S_i(\omega,\delta)$, then $\tilde{N}_t v \in S_i(\omega t,\delta)$ $(t \in I)$. Similarly for $U_i(\omega,\delta)$.

(c) $S_i(\omega,\delta) \supset \{v \in \pi^{-1}(\omega) \mid e^{-\lambda_i t}\tilde{N}_t v \to 0$ as $t \to \infty$ and $|\tilde{N}_t v|_* \leq \delta$ for all $t \geq 0\}$. Similarly for $U_i(\omega,\delta)$.

(d) $S_i(\omega,\delta)$, $U_i(\omega,\delta)$ depend Lipschitz continuously on ω. If $F(t,x)$ is C^r $(r \geq 1)$, they vary C^1-continuously. If $\lambda_i \leq 0$, $S_i(\omega,\delta)$ varies C^r-continuously; if $\lambda_i \geq 0$, $U_i(\omega,\delta)$ varies C^r-continuously.

(e) $S_i(\omega,\delta)$ is tangent at 0 to $\pi^{-1}(\omega) \cap V_1 \oplus \dots \oplus V_i$; $U_i(\omega,\delta)$ is tangent at 0 to $\pi^{-1}(\omega) \cap V_{i+1} \oplus \dots \oplus V_{n+1}$.

(f) $Q_i(\omega,\delta)$ contains $\{v \in \pi^{-1}(\omega) \mid |N_t v|_* \leq \delta,\ e^{-\lambda_i t} N_t v \to 0$ as $t \to \infty$, $e^{-\lambda_{i-1} t} \to 0$ as $t \to -\infty\}$.

(g) If $\lambda_i \leq 0$ and F is C^r $(r \geq 1)$, then $v \in S_i(\omega,\delta) \Longrightarrow \tilde{N}_t v$ is defined and in $S_i(\omega t,\delta)$ $(t \geq 0)$. Also, $S_i(\omega,\delta) = \{v \in \pi^{-1}(\omega) \mid e^{-\lambda_i t}|\tilde{N}_t v|_* \leq \delta$ for all $t \geq 0\}$. If $v \in S_i(\omega,\delta)$, then $|\tilde{N}_t v|_* \to 0$ as $t \to \infty$.

(h) If $\lambda_i \geq 0$, statements analogous to those in (g) hold for $t \leq 0$,

$t \to -\infty$.

Remark 2.22. Part (g) says that $S_i(\omega,\delta)$ is a genuine local stable manifold if $\lambda_i \leq 0$. Part (h) says $U_i(\omega,\delta)$ is a genuine local unstable manifold if $\lambda_i \geq 0$. That is, $\cup_{\omega\in\Omega} S_i(\omega,\delta)$ is positively invariant for equations $(2)_\omega$; $\cup_{\omega\in\Omega} U_i(\omega,\delta)$ is negatively invariant.

The proof of 2.21 uses the characterizations $S_i(\omega) = \text{graph}(H_1|\pi^{-1}(\omega))$, $U_i(\omega) = \text{graph}(H_2|\pi^{-1}(\omega))$, and properties of H_1 and H_2.

REFERENCES

1. H. Fenischel, Persistence and smoothness of invariant manifolds of flows, Indiana Univ. Math. J. 21(1971-72), 193-226.

2. M.J. Foster, Calculus on vector bundles, J. London Math. Soc. 11 (1975), 65-73.

3. M.J. Foster, Fibre derivatives and stable manifolds: a note, Bull. London Math. Soc. 8(1976), 286-288.

4. V. Guillemin, and A. Pollack, Differential Topology, Prentice-Hall, Inc., Englewood Cliffs, N.J., 1974.

5. P. Hartman, Ordinary Differential Equations, John Wiley and Sons, N.Y., 1964.

6. M. Hirsch and C. Pugh, Stable manifolds and hyperbolic sets, Proc. Symp. Pure Math., Vol. 14, A.M.S., Providence, R.I., 1970, 133-164.

7. M. Hirsch, C. Pugh and M. Shub, Invariant Manifolds, preprint.

8. M.C. Irwin, On the smoothness of the composition map, Quart. J. Math. Oxford 23(1972), 113-133.

9. J.L. Kelley, General Topology, D. Van Nostrand Co., Inc., Princeton, 1955.

10. R. Sacker, A perturbation theorem for invariant Riemannian manifolds, Proc. Symp. Univ. Puerto Rico, Academic Press, 1967, 43-54.

11. R. Sacker and G. Sell, Lifting properties in skew-product flows with applications to differential equations, Memoirs of A.M.S., #109.

12. G. Sell, Linear Differential Systems, Lecture Notes, Univ. of Minnesota, 1974.

13. G. Sell, The structure of a flow in the vicinity of an almost periodic solution, preprint.

UNIVERSITY OF SOUTHERN CALIFORNIA

LIFTING IN NON-ABELIAN (G,τ)-EXTENSIONS

by

H.B. Keynes* and D. Newton

§1. Introduction

In this note, we shall be concerned with extending the major lifting results in [5] to the case of a non-abelian group. In the previous paper, we examined the situation where (X,ψ) was a free (G,τ)-extension of (Y,η), G compact metric abelian, and considered perturbations of the system $(X,\psi) \xrightarrow{(G,\tau)} (Y,\eta)$ via τ-cocycles. Two conditions on (G,τ) were distinguished:

(A) For every $\varepsilon > 0$, there is an $N \geq 1$ such that if $n \geq N$ and $g_0,\ldots,g_{n-1} \in G$, then $\prod_{i=0}^{n-1} \tau^i(S_\varepsilon(g_i)) = G$ and,

(B) If $f \in C(Y,G)$ and $\varepsilon > 0$, there exists $\delta \geq 0$ such that if F is a finite subset of Y and $u: F \to G$ satisfies $d(f(y),u(y)) < \delta\,(y \in F)$, then there exists $v \in C(Y,G)$ with $u|_F = v|_F$ and $d(f(y),v(y)) < \varepsilon\,(y \in Y)$.

(G,τ) was called <u>admissible</u> if it satisfied (A) and (B). The major result shown was the following, under the assumption that all spaces involved were metric:

Theorem: <u>Let</u> $\pi: (X,\psi) \to (Y,\eta)$ <u>be a</u> (G,τ)-<u>extension</u>, Y <u>infinite with no isolated points and</u> (G,τ) <u>admissible</u>.

I. <u>If</u> (Y,η) <u>is point-transitive, then for almost all</u> $\varphi \in C(Y,G)$, (X,ψ_φ) <u>is point transitive</u>.

II. <u>If</u> (Y,η) <u>is minimal and</u> (G,τ) <u>distal, then for almost all</u> $\varphi \in C(Y,G)$, (X,ψ_φ) <u>is minimal</u>.

*Partially supported by National Science Foundation (MPS75-05250).

In this result, we have identified τ-cocycles with $C(Y,G)$. Finally, it was noted that if G is an n-torus, $1 \leq n \leq \infty$, then (G,τ) is admissible.

In Section 2, we note that the main theorem can be obtained in the case when G is non-abelian. We also prove that if G is a compact connected Lie group, then (G,τ) is admissible. This essentially is the full generalization of Ellis' original result for group extensions [1].

The final Section 3 is devoted to looking for admissible groups which are not Lie groups. It is noted that for a compact connected group with trivial center, admissibility is obtained, and partial results hold for ergodic automorphisms. We end with a few concluding remarks about other approaches to obtaining these results.

§2. The Lifting Theorem for a Compact Connected Lie Group

We start with a free (G,τ)-extension (X,ψ) of (Y,η), where G is not necessarily abelian and all spaces are metric. A τ-cocycle is defined the same way as in the abelian case, and the correspondence between τ-cocycles and $C(Y,G)$ still holds. We note however that given $\varphi \in C(Y,G)$, the perturbed flow (X,ψ_φ) is no longer a (G,τ)-extension of (Y,η), but simply an extension with fibre G. We do have:

Lemma 1: <u>If (G,τ) is distal, then (X,ψ_φ) is a distal extension of (Y,η).</u>

<u>Proof</u>: Let $\pi(x) = \pi(x_1)$, and $\lim_i \psi_\varphi^{n_i}(x) = \lim_i \psi_\varphi^{n_i}(x_1)$. Then $x_1 = gx$ for some g and, using notation of [5, Section 3], $\psi_\varphi^{n_i}(x_1) = \psi_\varphi^{n_i}(gx) = \overline{\varphi}(\pi x, n_i)\tau^{n_i}(g)\psi^{n_i}(x)$. Without loss of generality, assume $\psi^{n_i}(x) \to z$, $\varphi(\pi x, n_i) \to h$, and $\tau^{n_i}(g) \to k$. We then obtain $hz = hkz$, and thus $k = e$ by free action. This means that $\tau^{n_i}(g) \to e$ and

thus $g = e$, since τ is distal. So $x = x_1$, completing the proof.

It is direct to verify that the proof of [5, Theorem 3.13] holds for an arbitrary G. If (Y,η) is minimal and (G,τ) is distal, then π is a distal extension by Lemma 1, and so [5, Corollary 3.14] is obtained as before. We thus have shown

Theorem 2. _I and II of the main theorem are true for any arbitrary_ G _if_ (G,τ) _is admissible_.

We now turn to the question of when (G,τ) is admissible. Since G is compact metric, there exists a positive sequence $\varepsilon_i \to 0$ such that if $g \in G$, $gS_{\varepsilon_i}(e)g^{-1} = S_{\varepsilon_i}(e)$. Referring to the results in [5, Section 3], the condition in Lemma 3.2 holds for any of these ε_i's, and since for every $\varepsilon > 0$, $\varepsilon_i < \varepsilon$ for some i, Lemma 3.2 remains valid. Choosing an ε_i such that $S_{\varepsilon_i}(e)S_{\varepsilon_i}(g) \subset S_\varepsilon(g)$ $(g \in G)$ shows that Lemmas 3.3 and 3.4 remain valid. These comments yield the extended versions of Propositions 3.5 and 3.8.

Proposition 3. _Let_ G _be connected and_ τ _equicontinuous, or_ G _arbitrary and_ τ _ergodic_. _Then_ (G,τ) _satisfies_ (A).

We can now show the main result.

Theorem 4. _Let_ G _be a connected Lie group_. _Then_ (G,τ) _is admissible_.

Proof: Since Ellis shows [1, Proposition 2] that any compact connected Lie group satisfies (B), we need only show that (A) holds.

Let $\mathfrak{g}$ be the Lie algebra of G, and consider the Levi-Mal'cev decomposition $\mathfrak{g} = \mathfrak{R} + \mathfrak{S}$, $\mathfrak{R}$ solvable, $\mathfrak{S}$ semi-simple [7, pp. 256-257]. By the uniqueness of $\mathfrak{R}$, it follows that $\mathfrak{R}$ is $\hat{\tau}$-invariant, where $\hat{\tau}$ is the induced automorphism on $\mathfrak{g}$. This means that $G = RS$, where R is a connected, solvable, τ-invariant normal subgroup, and S is a connected semisimple subgroup (we can assume connected by using components of the identity). Since R is compact solvable, then R is abelian [7, p. 255], and we have the following diagram:

$$\begin{array}{c} (G,\tau) \\ (R,\tau)\ \Big\downarrow \\ (G/R,\tau) \end{array}$$

Consider the induced homomorphism $S \to G/R$. Since S is semi-simple, it follows [3] that if $\Sigma(S)$ is the dual group of S, and n is a positive integer, $\{\bar{\alpha} \in \Sigma(S) \mid \deg \alpha = n\}$ is finite ($\bar{\alpha}$ represents the class of α). Hence, the same property holds for $\Sigma(G/R)$. Pick $\bar{\alpha} \in \Sigma(G/R)$ and consider $\{\alpha\tau^p \mid p \text{ integer}\}$. Since $\deg \alpha\tau^p = \deg \alpha$, there exists a p with $\overline{\alpha\tau^p} = \bar{\alpha}$. By [4, Theorem 1.2], this implies that $(G/R,\tau)$ is equicontinuous. So $(G/R,\tau)$ satisfies (A) by Proposition 3. Since R is a torus, (R,τ) satisfies (A). By the extended version of Lemma 3.3, it follows that (G,τ) satisfies (A), which completes the proof.

Note that since R in the above tower is a finite-dimensional torus, the tower decomposes into a finite tower in the sense of [5], with a non-abelian equicontinuous base and connected fibers. It may have an ergodic piece, which will be an automorphism of a torus. If G itself is semi-simple, the above proof indicates that (G,τ) is equicontinuous. Actually, more is known: there is a positive integer r such that τ^r is an inner-automorphism. At any rate, we have:

Corollary 5. *Let* $\pi: (X,\psi) \to (Y,\eta)$ *be a* (G,τ) *extension with* Y *infinite*, G *connected semi-simple Lie group, and* (Y,η) *minimal. Then for almost all* $\varphi \in C(Y,G)$, (X,ψ_φ) *is minimal.*

§3. Extensions to Non-Lie Groups and Concluding Remarks

We now examine when G can be admissible without being a Lie group. Clearly, (B) holds for any finite group. A simple modification of [5, Proposition 3.11] shows that if (B) holds for G_i $(i \geq 1)$, then it holds for $X_{i=1}^{\infty} G_i$. This means that (B) holds for any countable product of groups in which each factor is either a finite group or a compact connected Lie group.

Recall that a Bernoulli group automorphism on G is an automorphism τ such that (G,τ) is isomorphic to $X_{-\infty}^{\infty} H$ with the shift map, where H is compact metric. Now let G be a connected group without center. Then it is known [6, p. 115] that (G,τ) is isomorphic to $(G_1 \times G_2, \tau_1 \times \tau_2)$, where (G_1,τ_1) is Bernoulli and the state group H is a connected simple Lie group and (G_2,τ_2) is a countable direct product of automorphisms of connected semi-simple Lie groups. Since τ_1 is ergodic, (G,τ_1) satisfies (A) by Proposition 3 and (B) by the above comments. Since (G_2,τ_2) is a product of equicontinuous automorphisms, it is equicontinuous and hence satisfies (A) also by Proposition 3, and (B) by the above comments. Since (A) is preserved under products, we thus have shown:

Theorem 6. <u>Let</u> G <u>be a compact connected group with trivial center</u>. <u>Then</u> (G,τ) <u>is admissible</u>.

If (G,τ) is ergodic, we have by Proposition 3 that such a (G,τ) is admissible whenever G satisfies (B). Even without (B), a related result holds:

Proposition 7. <u>Let</u> $\pi\colon (X,\psi) \to (Y,\eta)$ <u>be a</u> (G,τ)-<u>extension, where</u> Y <u>is infinite</u>, (Y,η) <u>minimal and</u> (G,τ) <u>ergodic</u>. <u>Then</u> (Y,ψ) <u>is point transitive</u>.

<u>Proof</u>: Let μ be an ergodic measure on Y. Since Y is minimal, μ is supported. Thus, the Haar Lift $\tilde{\mu}$ is ergodic [4, Corollary 2.2]

and supported, and so (X,ψ) is point-transitive.

We conclude with a few remarks on other possible extensions of the results in [5].

1. Following some suggestions of Weiss and Furstenberg concerning an alternate proof of (A) for abelian G, the following measure-theoretic version of (A) seems plausible: Let m be Haar measure on G, with G abelian. Let $\varepsilon > 0$, and (S_n) a sequence of sets with $m(S_n) > \varepsilon$ for every n. Then there exists N such that if $n \geq N$, $S_1 S_2 \ldots S_n = G$.

2. Ellis has developed [2] an algebraic theory of group extensions and their perturbations. Some initial investigations indicate that in the case of (Y,η) minimal, there may be some relationship between (G,τ)-extensions and the perturbed flows $\mathrm{per}(\beta,\sigma)$.

REFERENCES

1. R. Ellis, The construction of minimal discrete flows, Amer. J. Math. 87(1965), 564-574.

2. R. Ellis, Cocycles in topological dynamics, Topology, to appear.

3. Harish-Chandra, On representations of Lie algebras, Ann. of Math. (2)50(1949), 900-915.

4. H. Keynes and D. Newton, Ergodicity in (G,σ)-extensions, submitted.

5. H. Keynes and D. Newton, Minimal (G,τ)-extensions, Pacific J. Math., to appear.

6. R.K. Thomas, Metric properties of transformations of G-spaces, Trans. Amer. Math. Soc. 160(1971), 103-117.

7. D.P. Zelobenko, Compact Lie Groups and Their Representations, AMS Translations, Vol. 40, 1973.

H.B. Keynes
UNIVERSITY OF MINNESOTA

D. Newton
UNIVERSITY OF SUSSEX

RECIPE MINIMAL SETS

by

Nelson G. Markley[1] and Michael E. Paul

In a paper on asymptoticity Baum [1] gave an example of an almost periodic point in $\{0,1\}^{\mathbf{Z}\oplus\mathbf{Z}}$ under the obvious $\mathbf{Z}\oplus\mathbf{Z}$ action ($\mathbf{Z}$ denotes the integers) which has properties like the Morse sequence in $\{0,1\}^{\mathbf{Z}}$. He constructed the point by using only the Morse sequence and its dual as vertical sequences and using the Morse sequence to arrange them in the horizontal direction. Specifically let μ denote the Morse sequence and $\bar{\mu}$ its dual and define $z(m,n) = \mu(n)$ or $\bar{\mu}(n)$ for all n according as $\mu(m) = 0$ or 1. Is this a singular example or is the class of almost periodic points in $\{0,1\}^{\mathbf{Z}\oplus\mathbf{Z}}$ with only a finite number of vertical sequences an interesting one to study?

In this note we will expose the fundamental structure of such "sequences" and their orbit closures. Although they do provide an easy source of examples exhibiting a variety of properties, they are in a definite sense the least interesting almost periodic points in $\{0,1\}^{\mathbf{Z}\oplus\mathbf{Z}}$. It turns out that modulo a finite group they are cartesian products of minimal sets from $\{0,1\}^{\mathbf{Z}}$.

§1. Definitions and Fundamental Theorems

Let $(\Omega_1(p),\sigma)$ denote the usual full shift system on the p symbols $0, 1,\ldots,p-1$. Set $\Omega_2(p) = \{0, 1,\ldots,p-1\}^{\mathbf{Z}\oplus\mathbf{Z}}$ with the product topology coming from the discrete topology on $\{0, 1,\ldots,p-1\}$. Define σ on $\Omega_2(p)$ by $\sigma(k,\ell)x(m,n) = x(m+k, n+\ell)$. Then $(\Omega_2(p),\sigma)$ will be called the full shift system of rank 2 on p symbols. The orbit of a point z in $\Omega_i(p)$ will be denoted by $\mathcal{O}(z)$.

[1]Supported by National Science Foundation (MPS75-07078)

Let z' be an almost periodic point in $\Omega_2(p)$ for which there exists a finite number of distinct points $y'_0, y'_1, \ldots, y'_{q-1}$ in $\Omega_1(p)$ such that given $m \in \mathbf{Z}$ there exists $i_m \in \{0,\ldots,q-1\}$ satisfying $z'(m,n) = y'_{i_m}(n)$ for all $n \in \mathbf{Z}$. The set of points $\{y'_0,\ldots,y'_{q-1}\}$ will be called the ingredients of z'. We can and will assume that each $k \in \{0,\ldots,q-1\}$ is some i_m. It is easy to see that every point in $\overline{O(z')}$ has this property and q is constant on $\overline{O(z')}$. Let $x' \in \Omega_1(q)$ be defined by $x'(m) = i_m$ and let $\bar{y} = (y'_0,\ldots, y'_{q-1}) \in [\Omega_1(p)]^q$. The point x' will be called a recipe for obtaining z' from its ingredients, we will write $z' = \bar{y}'_{x'}$, and we will call $\overline{O(z')}$ a recipe minimal set. Let $X = \overline{O(x')}$, let $Y = \overline{O(\bar{y}')}$, and form $(X \times Y, \mathbf{Z} \oplus \mathbf{Z})$ in the obvious way. Note that x', X, and Y depend upon the indexing of the ingredients of z'. Define $\rho\colon X \times Y \to \Omega_2(p)$ by $\rho(x,\bar{y}) = \bar{y}_x$. It is easy to check that ρ is a homomorphism whose image is $\overline{O(z')}$.

Theorem 1. <u>In this situation</u> x' <u>and</u> $\bar{y}'$ <u>are almost periodic points of</u> $\Omega_1(q)$ <u>and</u> $[\Omega_1(p)]^q$ <u>respectively and</u> $\rho\colon X \times Y \to \overline{O(z')}$ <u>is a finite group extension</u>. <u>If</u> $\overline{O(y_i)}$ <u>are distinct, then</u> ρ <u>is an isomorphism of</u> $X \times Y$ <u>onto</u> $\overline{O(z')}$. <u>Moreover, if there exist minimal flows</u> $(X_1,\mathbf{Z})$ <u>and</u> $(Y_1,\mathbf{Z})$ <u>such that</u> $\overline{O(z')}$ <u>is isomorphic to</u> $(X_1 \times Y_1, \mathbf{Z} \oplus \mathbf{Z})$, <u>then</u> $(X_1,\mathbf{Z})$ <u>and</u> $(Y_1,\mathbf{Z})$ <u>are isomorphic to</u> $(X,\mathbf{Z})$ <u>and</u> $(Y,\mathbf{Z})$ <u>respectively</u>.

<u>Proof</u>. Let M be a minimal subset of $(X \times Y, \mathbf{Z} \oplus \mathbf{Z})$. Then $\rho(M) = \overline{O(z')}$ and $M = \overline{O(x)} \times \overline{O(\bar{y})}$ where x and $\bar{y}$ are almost periodic points and $\bar{y}_x = z' = \bar{y}'_{x'}$. In addition, $\pi(x(i)) = x'(i)$ defines a permutation on $\{0,\ldots,q-1\}$, and it naturally defines automorphisms of $\Omega_1(q)$ and $[\Omega_1(p)]^q$ which map x and y to x' and y'. For the last statement one first notes that ρ restricted to $X \times \{y\}$ is one-to-one. Letting $\varphi\colon \overline{O(z')} \to X_1 \times Y_1$ be an isomorphism one then

checks that $\varphi\circ\rho(\mathcal{O}(x) \times \{y\}) = \mathcal{O}(x_1) \times \{y_1\}$ for some (x_1,y_1). It follows that $\varphi\circ\rho(\overline{\mathcal{O}(x)} \times \{y\}) = X_1 \times \{y_1\}$ and mapping x to the first coordinate of $\varphi\circ\rho(x,y)$ is the desired isomorphism. Finally, Y is handled in the same way.

Notice that the last part of Theorem 1 does not assert that ρ is one-to-one. It does imply that if $\overline{\mathcal{O}(z')}$ has some dynamical behavior which $X \times Y$ does not have, then $\overline{\mathcal{O}(z')}$ is not isomorphic to a product of two minimal sets.

Theorem 2. *Let* M *be a minimal subset of* $\Omega_2(p)$. *Then* M *is a recipe minimal set if and only if* M *is a homomorphic image of some* $(X \times Y, \mathbb{Z} \oplus \mathbb{Z})$ *where* $(X,\mathbb{Z})$ *and* $(Y,\mathbb{Z})$ *are symbolic minimal sets.*

Proof. The "only if" part follows from Theorem 1. Suppose φ is a homomorphism of $(X \times Y, \mathbb{Z} \oplus \mathbb{Z})$ onto M. Let $U_i = \varphi^{-1}(\{z\colon z(0,0) = i\})$. Then each U_i is open and closed. Let A_k and B_k denote the $2k+1$-blocks occurring in X and Y respectively. There exists k' such that for all $\alpha \in A_{k'}$ and $\beta \in B_{k'}$ the set $\{(x,y)\colon x(-k'),\ldots,x(k') = \alpha$ and $y(-k'),\ldots,y(k') = \beta\}$ is contained in some U_i. This defines a natural map $f\colon A_{k'} \times B_{k'} \to$ symbols in M. Set $f_\infty(x,y)(m,n) = f(\alpha,\beta)$ where $\alpha = x(m),\ldots,x(m+2k')$ and $\beta = y(n),\ldots,y(n+2k')$. It is easy to check that $f_\infty = \varphi\circ(\sigma^{k'} \times \sigma^{k'})$. Fixing $x \in X$ and $y \in Y$ observe that $f_\infty(x,y)(m_0,n) = f_\infty(x,y)(m_1,n)\,\forall n$ if $x(m_0),\ldots,x(m_0+2k') = x(m_1),\ldots,x(m_1+2k')$. Since $A_{k'}$ is finite, only a finite number of different columns occur in $f_\infty(x,y)$ and we are done.

Note the last step of this prooof works for rows as well as columns which proves the following:

Corollary 1. *Let* M *be a recipe minimal set in* $\Omega_2(p)$ *and let* $z \in M$. *Then there exist* $x_1,\ldots,x_r \in \Omega_1(p)$ *such that given* n *there is an* i *satisfying* $z(m,n) = x_i(m)$ *for all* m.

Corollary 2. Let M and M' be minimal subsets of $(\Omega_2(p),\sigma)$ and let $\varphi: M \to M'$ be a homomorphism. If M is a recipe minimal set, then so is M'.

Corollary 3. Let $(X,\mathbb{Z})$ and $(Y,\mathbb{Z})$ be symbolic minimal sets. Then there exists a recipe minimal set which is isomorphic to $(X \times Y, \mathbb{Z} \oplus \mathbb{Z})$.

The preceding can easily be extended to $\Omega_N(p)$. To see this suppose $m \in \mathbb{Z}^s$ and $n \in \mathbb{Z}^t$ where $s + t = N$. Then with each $y_i' \in \Omega_t(p)$ the formula $z'(m,n) = y_i(n)$ for all $n \in \mathbb{Z}^t$ makes sense and $x'(m) = \dot{i}_m$ is a point in $\Omega_s(q)$. In this setting all the results in this section are still valid. The proofs are the same and only the technical details in extending the proof of Theorem 2 are more complicated.

§2. Dynamical Properties

Let $(X,\mathbb{Z})$ and $(Y,\mathbb{Z})$ be minimal transformation groups on compact metric spaces. As in the previous section form $(X \times Y, \mathbb{Z} \oplus \mathbb{Z})$ in the obvious way. The following three lemmas are direct consequences of the relevant definitions:

Lemma 1. Let (x,y) and (x',y') be points in $X \times Y$. Then (x,y) and (x',y') are proximal {regionally proximal} if and only if (x,x') and (y,y') are in the proximal relation {regionally proximal relation} for $(X,\mathbb{Z})$ and $(Y,\mathbb{Z})$ respectively.

Lemma 2. Let $(x,y) \in X \times Y$. Then (x,y) is a distal point {almost automorphic point} {regularly almost periodic point} {isochronous point} if and only if x and y are distal points {almost automorphic points} {regularly almost periodic points} {isochronous points} of $(X,\mathbb{Z})$ and $(Y,\mathbb{Z})$ respectively.

Lemma 3. <u>The transformation group</u> $(X \times Y, \mathbf{Z} \oplus \mathbf{Z})$ <u>is topologically weak mixing {topologically strong mixing} if and only if both</u> $(X,\mathbf{Z})$ <u>and</u> $(Y,\mathbf{Z})$ <u>are topologically weak mixing {topologically stong mixing}</u>.

Lemma 4. <u>The transformation group</u> $(X \times Y, \mathbf{Z} \oplus \mathbf{Z})$ <u>is uniquely ergodic if and only if</u> $(X,\mathbf{Z})$ <u>and</u> $(Y,\mathbf{Z})$ <u>are uniquely ergodic.</u>

<u>Proof</u>. Assume $(X,\mathbf{Z})$ and $(Y,\mathbf{Z})$ are uniquely ergodic with unique invariant measures μ and υ. Let λ be an invariant measure on $X \times Y$. Clearly $\lambda(A \times Y) = \mu(A)$ for any Borel set A in X. Choose A such that $\mu(A) \neq 0$. Then similarly $\lambda(A \times B)/\lambda(A \times Y) = \upsilon(B)$ for all Borel sets B in Y. Thus if $\mu(A) > 0$ we have $\lambda(A \times B) = \mu(A)\upsilon(B)$. If $\mu(A) = 0$, then $\lambda(A \times Y) = 0$, $\lambda(A \times B) = 0$, and $\lambda(A \times B) = \mu(A)\upsilon(B)$. Therefore, $\lambda = \mu \times \upsilon$. The other half is standard because the projection on X is a homomorphism if one trivially extends the $\mathbf{Z}$ action to a $\mathbf{Z} \oplus \mathbf{Z}$ action on X.

Throughout the rest of this section M will be a recipe minimal set. Pick $z' \in M$ and let $\bar{y}' = (y_0,\ldots,y_{q-1})$ be the ingredients of z' written down in some order. This determines the recipe x' such that $z' = \bar{y}'_{x'}$. Let $X = \overline{\mathcal{O}(x')}$, $Y = \overline{\mathcal{O}(\bar{y}')}$, and $\rho(x,\bar{y}) = \bar{y}_x$. This notation will also be fixed for the rest of this section.

Theorem 3. <u>The recipe minimal set</u> M <u>is almost automorphic {point distal} if and only if both</u> $(X,\mathbf{Z})$ <u>and</u> $(Y,\mathbf{Z})$ <u>are almost automorphic {point distal}</u>.

<u>Proof</u>. The only part which is not obvious from the previous results is $(X \times Y, \mathbf{Z} \oplus \mathbf{Z})$ is almost automorphic when M is. Suppose $\bar{y}_x$ is almost automorphic and $(x,\bar{y})$ is not. Thus either x or $\bar{y}$ is not almost automorphic, suppose it is x. Then there exists m_i such that $\sigma^{m_i}(x) \to x_0$ and $\sigma^{-n_i}(x_0) \to x'' \neq x$. Using the sequence $(m_i,0)$ with $\bar{y}_x$ we get the contradiction $\bar{y}_x = \bar{y}_{x''}$. (Remember we don't

allow repeats in $\bar{y}$.) The argument for $\bar{\bar{y}}$ is similar.

Theorem 4. <u>The recipe minimal set</u> M <u>is topologically weak mixing</u> {<u>topologically strong mixing</u>} {<u>uniquely ergodic</u>} <u>if</u> $(X,\mathbb{Z})$ <u>and</u> $(Y,\mathbb{Z})$ <u>have this property.</u>

<u>Proof</u>. Use Lemmas 3 and 4 and note that these properties are preserved under homomorphism.

If the homomorphism ρ is an isomorphism, then the maximal equicontinuous factor of the recipe minimal set M is $(G \times H, \mathbb{Z} \oplus \mathbb{Z})$ where $(G,\mathbb{Z})$ and $(H,\mathbb{Z})$ are the maximal equicontinuous factors of X and Y. This follows from Lemma 1 because the regionally proximal relation coincides with the equicontinuous structure relation for abelian groups. It is easy to show that a finite group extension induces a homomorphism between the maximal equicontinuous factors which is a group extension under the induced action of the finite group. Thus if Π is the finite group such that $(X \times Y)/\Pi$ is M, then the maximal equicontinuous factor of M is $(G \times H)/\tilde{\Pi}$ where $\tilde{\Pi}$ is the group of homomorphisms induced on $G \times H$ by Π. Let $\pi \in \Pi$. If $\pi(x,y)$ and (x,y) are regionally proximal, then $\tilde{\pi}$ has a fixed point and must be the identity. Thus in general $\tilde{\Pi}$ is a smaller group than Π; in fact, $\tilde{\Pi}$ can be trivial. (See Example 1 in the next section). On the other hand, we have an example showing that it may equal $G \times H$.

Given the almost periodic behavior of the rows and columns of points in M, it is natural to ask about the behavior in other directions. Example 1 in the next section shows that they need not be almost periodic. However, as Theorem 5 shows, in many cases all such sequences are almost periodic in $\Omega_1(p)$.

Theorem 5. <u>If</u> $(X,\mathbb{Z})$ <u>and</u> $(Y,\mathbb{Z})$ <u>are disjoint and at least one of them is point distal, then for any</u> $p, q, k \in \mathbb{Z}$ <u>and</u> $z \in M$, <u>the</u>

<u>sequence</u> $w(n) = z(pn, qn+k)$ <u>is an almost periodic point in</u> $\Omega_1(p)$.

<u>Proof</u>. We may as well assume neither p nor q is zero. It is easy to see that $w(n)$ is almost periodic if $(x,\bar{y})$ is almost periodic with respect to $\sigma^p \times \sigma^q$ where $z = \bar{y}_x$. (Here the second σ stands for the appropriate cartesian product of shifts restricted to Y.) Thus if suffices to show that $\overline{O(x,\sigma^p)}$ and $\overline{O(\bar{y},\sigma^q)}$ are disjoint. From the point distal hypothesis we know this is equivalent to showing they have no common eigenvalues [2]. Since the eigenvalues of ψ^N are the Nth powers of the eigenvalues of ψ, we are done.

§3. Examples

We now present three examples which show the general behavior of recipe minimal flows. All are recipe flows which are not isomorphic to product flows. The first example, originally appearing in Baum [1], is discussed along with a particular extension and factor. The second example, while very similar to the first, has in addition the property that all its directions are almost periodic. The third example uses recipes and ingredients quite different from the first two, and provides an example of a weakly mixing recipe flow constructed from ingredients which are not weakly mixing. The section concludes with an example of a minimal flow which cannot be a recipe flow, and hence cannot be a factor of any $(X \times Y, \mathbb{Z} \oplus \mathbb{Z})$.

Example 1. Let μ denote the Morse bisequence, let $(X,\mathbb{Z}) = \overline{(O(\mu)},Z)$ and let $(Y,Z) = (\overline{O((\mu,\bar{\mu}))}, \mathbb{Z})$. (Here $\bar{\mu}$ denotes the dual of μ obtained by switching 0's and 1's.) Now set $(M, \mathbb{Z} \oplus \mathbb{Z}) = (\rho(X \times Y), \mathbb{Z} \oplus \mathbb{Z})$, where ρ is the homomorphism described in the preceding sections. The flow $(M, \mathbb{Z} \oplus Z)$ is Baum's example. The block map $x_1 + x_2$ gives an exactly 2-to-1 homomorphism from $(\overline{O(\mu)},\mathbb{Z})$ onto an almost automorphic flow $(\overline{O(\upsilon)}, \mathbb{Z})$. To extend this to a homomorphism of M, code the 1×2 blocks appearing in M and define f_∞ by:

$$f(ab) = c$$

where $a = \begin{pmatrix} a_2 \\ a_1 \end{pmatrix}$, $b = \begin{pmatrix} b_2 \\ b_1 \end{pmatrix}$, and $c = \begin{pmatrix} c_2 = a_1 + a_2 \\ c_1 = a_1 + b_1 \end{pmatrix}$.

We now have the following situation -

$$(X \times Y, \mathbb{Z} \oplus \mathbb{Z}) \xrightarrow[\text{2-to-1}]{\rho} (M, \mathbb{Z} \oplus \mathbb{Z}) \xrightarrow[\text{2-to-1}]{f_\infty} (f_\infty(M), \mathbb{Z} \oplus \mathbb{Z}),$$

where both $X \times Y$ and M are $\mathbb{Z}_2$ group extensions.

Proposition 1. <u>The mapping</u> $\rho(x,y) = y_x$ <u>is exactly</u> 2-<u>to</u>-1; $\rho(x_1,y_1) = \rho(x_2,y_2)$ <u>iff</u> $x_1 = \bar{x}_2$ <u>and</u> $y_1 = \bar{y}_2$. <u>In particular</u>, $\rho(\mu,(\mu,\bar{\mu})) = \rho(\bar{\mu},(\bar{\mu},\mu))$.

Proposition 2. <u>The recipe flow</u> $(M, \mathbb{Z} \oplus \mathbb{Z})$ <u>is not isomorphic to a product</u>.

<u>Proof</u>. The coalescence of the Morse minimal set and Proposition 5 below prove this proposition. We provide an alternate proof because the technique is useful when information on coalescence is not available. Note that $(X \times Y, \mathbb{Z} \oplus \mathbb{Z})$ has a regionally proximal cell of cardinality 16, while the largest regionally proximal cell of $(M, \mathbb{Z} \oplus \mathbb{Z})$ is of cardinality 8. Thus M and $X \times Y$ are not isomorphic, and so by Theorem 1 it follows that $(M, \mathbb{Z} \oplus \mathbb{Z})$ cannot be isomorphic to a product.

Proposition 3. <u>The recipe flow</u> $(f_\infty(M), \mathbb{Z} \oplus \mathbb{Z})$ <u>is isomorphic to a product</u>.

<u>Proof</u>. Recall that the block map $x_1 + x_2$ gives a homomorphism $h_1: (\overline{O(\mu)}, \mathbb{Z}) \to (\overline{O(\upsilon)}, \mathbb{Z})$. Consider the diagram -

$$\begin{array}{ccc} (X \times Y,\ \mathbb{Z} \oplus \mathbb{Z}) & \xrightarrow{f_\infty \circ \rho} & (f\ (M),\ \mathbb{Z} \oplus \mathbb{Z}) \\ \downarrow h & & \\ (\overline{\mathcal{O}(\upsilon)} \times \overline{\mathcal{O}(\upsilon)},\ \mathbb{Z} \oplus \mathbb{Z}), & & \end{array}$$

where $h(x,y) = (h_1(x),h_1(y))$. It follows immediately that $f_\infty \circ \rho$ identifies precisely the same points in $X \times Y$ as h does, and this proves the proposition.

Proposition 4. <u>The recipe flow $(M, \mathbb{Z} \oplus \mathbb{Z})$ has the following properties:</u>

(a) <u>It is uniquely ergodic.</u>

(b) <u>It is point distal but not almost automorphic.</u>

(c) <u>It has maximal equicontinuous factor equal to $(\mathbb{Z}(2) \times \mathbb{Z}(2), 1 \oplus 1)$. (Here $1 \oplus 1$ denotes the $\mathbb{Z} \oplus \mathbb{Z}$ action obtained by letting 1 act independently on each factor by translation.)</u>

(d) <u>The induced maps $\tilde{\rho}$ and $\tilde{f}_\infty$ defined on $\mathbb{Z}(2) \times \mathbb{Z}(2)$ are both the identity map.</u>

<u>Proof</u>. Use the results of section 2.

Remark. Consider the point $w \in M$, $w = \rho(\tilde{\mu}, (\mu,\bar{\mu}))$, where

$$\tilde{\mu}(n) = \begin{cases} \mu(n), & n \geq 0 \\ \overline{\mu(n)}, & n < 0 \end{cases}. \quad \text{Then} \quad w(n,n) = \begin{cases} 0, & n \geq 0 \\ 1, & n < 0 \end{cases}.$$

Thus the recipe flow M contains points with directions that are not almost periodic. It also follows that $(\tilde{\mu}, (\mu,\bar{\mu})) \in X \times Y$ must be a non-almost periodic under the action of $\sigma \times \sigma$.

Example 2. Before presenting the second example, it will be useful to introduce Proposition 5, which gives conditions under which a recipe flow cannot be a product.

Lemma 5. *Let* $h: (X_1 \times Y_1, \mathbb{Z} \oplus \mathbb{Z}) \to (X_2 \times Y_2, \mathbb{Z} \oplus \mathbb{Z})$ *be a homomorphism of flows, where* X_1 *and* Y_1 *are each point transitive. Then there exist homomorphisms* h_1 *and* h_2 *of* $(X_1,\mathbb{Z})$ *and* $(Y_1,\mathbb{Z})$ *such that* $h(x,y) = (h_1(x),h_2(y))$ *for all* $(x,y) \in X_1 \times Y_1$. *Thus, in particular, if* $(X_1,\mathbb{Z})$ *and* $(Y_1,\mathbb{Z})$ *are each coalescent, then* $(X_1 \times Y_1, \mathbb{Z} \oplus \mathbb{Z})$ *is coalescent.*

Proof. Write $h(x,y) = (h_1(x,y), h_2(x,y))$. Then:

$$h((n,0)(x,y)) = (h_1(nx,y), h_2(nx,y)) = (n,0)h(x,y) = (nh_1(x,y), h_2(x,y)).$$

Thus $h_2(nx,y) = h_2(x,y)$ and so h_2 depends only on y. Similarly h_1 depends only on x.

Proposition 5. *Suppose* $(X_1,\mathbb{Z})$ *and* $(Y_1,\mathbb{Z})$ *are each coalescent and the recipe map* $\rho: X_1 \times Y_1 \to \rho(X_1 \times Y_1)$ *is not* 1-to-1. *Then the recipe flow* $(\rho(X_1 \times Y_1), \mathbb{Z} \oplus \mathbb{Z})$ *cannot be isomorphic to a product.*

Proof. If $\rho(X_1 \times Y_1)$ were a product, then Theorem 1 would imply the existence of an isomorphism $\varphi: \rho(X_1 \times Y_1) \to X_1 \times Y_1$. But then $\varphi \circ \rho$ would be an endomorphism of $X_1 \times Y_1$ which was not 1-to-1, violating the preceding lemma.

Now let $(X,\mathbb{Z})$ be a dual invariant coalescent minimal symbolic flow with the circle and irrational rotation as maximal equicontinuous factor. To construct such a flow, let $K = \mathbb{R}/\mathbb{Z}$, $A = [0,\frac{1}{2}]$, γ = any irrational number, and define the bisequence x by the formula

$$x(n) = \begin{cases} 0 & \text{if } n\gamma \notin A \\ 1 & \text{if } n\gamma \in A \end{cases}.$$

Then set $(X,\mathbb{Z}) = (\overline{\mathcal{O}(x)}, \mathbb{Z})$. Now let $(Y,\mathbb{Z}) = (\overline{\mathcal{O}((\mu,\bar{\mu}))}, \mathbb{Z})$ where as usual μ denotes the Morse bisequence.

Proposition 6. <u>The recipe flow</u> $(\rho(X \times Y), \mathbb{Z} \oplus \mathbb{Z})$ <u>has the following properties</u>:

(a) <u>It is uniquely ergodic</u>.

(b) <u>It is point distal but not almost automorphic</u>.

(c) <u>It has maximal equicontinuous factor equal to</u> $(K \times \mathbb{Z}(2), 2\gamma \oplus 1)$.

(d) <u>The induced map</u> $\tilde{\rho}: K \times \mathbb{Z}(2) \to K \times \mathbb{Z}(2)$ <u>is</u> $\tilde{\rho}(k,z) = (2k,z)$.

(e) <u>It is not a product</u>.

(f) <u>All the points of</u> $\rho(X \times Y)$ <u>have all their directions being almost periodic</u>.

<u>Proof</u>. Parts (a), (b), (c), (d), and (f) are consequences of section 2. In particular, part (f) follows from Theorem 5. Part (e) follows from Proposition 5 of this section along with the well known coalescence of the Morse minimal set.

Example 3. Let $(X,\mathbb{Z})$ be a weakly mixing symbolic minimal flow which is dual invariant. The skew-product construction discussed in Peleg [3, p.335-336] has this property. Now consider $(X \times \mathbb{Z}_2, \mathbb{Z})$; it clearly has an automorphism ψ of period two (it is minimal because weakly mixing flows are disjoint from distal flows). Code any point $y \in X \times \mathbb{Z}_2$ so that it becomes a symbolic bisequence, and define $Y = \overline{\mathcal{O}((y,\psi(y))}$. Then by the results of section 2 the product flow $(X \times Y, \mathbb{Z} \oplus \mathbb{Z})$ has maximal equicontinuous factor $(0 \times \mathbb{Z}_2, 0 \oplus 1)$, while the recipe flow $(\rho(X \times Y), \mathbb{Z} \oplus \mathbb{Z})$ has trivial maximal equicontinuous factor. This also implies by Theorem 1 that this recipe flow is not isomorphic to a product.

Example 4. We now describe a regularly almost periodic point ω such that $(\overline{\mathcal{O}(\omega)}, \mathbb{Z} \oplus \mathbb{Z})$ is neither recipe nor isomorphic to $(X, \mathbb{Z} \oplus \mathbb{Z})$ where $(X, \mathbb{Z} \oplus 0)$ is minimal. Let $z \in \Omega_1(2)$ be a regularly almost periodic point such that the 2-block 00 does not appear in z. Then there is a partition of the integers $\mathbb{Z} = \bigcup_{i=1}^{\infty} P_i$ into disjoint arithmetic progressions such that $z(\ell) = z(m)$ if ℓ and m are both in the same P_i. Think of z as being the y-axis of w, i.e. $\omega(0,n) = z(n)$ for all integers n. Now fill in the horizontal lines of ω with periodic points from $\Omega_1(2)$ so that

(a) given ℓ and m is the same P_i, $\omega(j,\ell) = \omega(j,m)$ for all integers j

(b) all the periodic points from $\Omega_1(2)$ get used

(c) $\omega(1,0) = \omega(1,1) = 0$.

It follows immediately from this construction that ω is a regularly almost periodic point in $\Omega_2(2)$. Condition (b) guarantees that $(\overline{\mathcal{O}(\omega)}, \mathbb{Z} \oplus 0)$ is not minimal, and condition (c) along with the fact that the y-axis of ω misses the block 00 implies that $(\overline{\mathcal{O}(\omega)}, 0 \oplus \mathbb{Z})$ is not minimal. That $(\overline{\mathcal{O}(\omega)}, \mathbb{Z} \oplus \mathbb{Z})$ is not a recipe flow follows from the fact that ω has infinitely many distinct horizontal lines. Indeed, $\overline{\mathcal{O}(\omega)}$ has the property that every point from $\Omega_1(2)$ appears as a horizontal line in some point of $\overline{\mathcal{O}(\omega)}$.

BIBLIOGRAPHY

1. Baum, J.D., Asymptoticity in topological dynamics, Trans. Amer. Math. Soc. 77(1954), 506-519.

2. Keynes, H.B., Disjointness in transformation groups, Proc. Amer. Math. Soc. 36(1972), 253-259.

3. Peleg, R., Some extensions of weakly mixing flows, Israel J. Math. 9(1971), 330-331.

UNIVERSITY OF MARYLAND

LARGE SETS OF ENDOMORPHISMS AND OF g-MEASURES

by

Marion Rachel Palmer, William Parry and Peter Walters

The notes are the outcome of an unfulfilled attempt to prove that, in a reasonable sense, most endomorphisms of a Lebesgue space $(X,\mathcal{B},m)$ can be classified by a countable number of invariants. We attempted to do this despite Feldman's negative warnings in [2] concerning automorphisms; and in fact the invariants we had in mind will not classify. The complete metric topology we adopt for the space $E(X)$ of endomorphisms is stronger than the usual weak operator topology since we considered the latter to be inappropriate. In fact the set $A(X)$ of automorphisms of X is dense in $E(X)$ in the weak topology. This is one basic reason for adopting what we call the strong adjoint topology. In this topology $A(X)$ is a closed nowhere dense subset of $E(X)$. Another reason for adopting the strong adjoint topology is that certain conditional expectation operators are directly related to this topology.

With the strong adjoint topology the set of exact automorphisms is a dense G_δ in $E(X)$ whereas with the weak topology even the strong-mixing endomorphisms are a set of first category. We shall also show that the set of exact Markov endomorphisms and even the 'irregular' exact Markov endomorphisms are dense in $E(X)$ with respect to the strong adjoint topology. An endomorphism T is irregular if its information function $I(\mathcal{B}|T^{-1}\mathcal{B})$ generates the full σ-algebra $\mathcal{B}$. If T is irregular then it is characterised by a countable number of invariants. Contrary to our expectations we have shown that irregular endomorphisms form a set of first category; in fact $I(\mathcal{B}|T^{-1}\mathcal{B}) = \infty$ a.e. is the general case.

Due to this failure we turned to an analogous problem. We study the set M_g of so called g-measures (defined with respect to an aperiodic shift of finite type) with a natural topology stronger than the weak*-topology. In this topology the set of g-measures making the shift on exact endomorphism form a dense G_δ and the Markov measures are dense. The irregular g-measures form an open dense set and can be characterised by a countable number of invariants. Also most g's have a unique g-measure, but we are unable to decide if all g's have a unique g-measure. (See [4].)

1. The Space $\mathcal{E}(X)$

Let $(X,\mathcal{B},m)$ be a non-atomic Lebesgue space, i.e., a probability space isomorphic to the unit interval with Lebesgue measurable sets and Lebesgue measure. A measure-preserving transformation of $(X,\mathcal{B},m)$ is called an <u>endomorphism</u> of $(X,\mathcal{B},m)$. The space (semi-group) of all endomorphisms of $(X,\mathcal{B},m)$ will be denoted by $\mathcal{E}(X)$. The subset (subgroup) of $\mathcal{E}(X)$ consisting of all invertible endomorphisms (<u>automorphisms</u>) will be denoted by $\mathcal{A}(X)$. The group $\mathcal{A}(X)$ is usually endowed with the weak operator topology inherited from the weak topology on the group $U(L^2(X))$ of unitary operators on $L^2(X)$ by the injection $\mathcal{A}(X) \ni T \to U_T \in U(L^2(X))$ where $U_T f = f\circ T$. ([3].) The injection $\mathcal{E}(X) \ni T \to U_T$ associates an isometry U_T of $L^2(X)$ to each T. The weak and strong topologies coincide on the set of isometries of $L^2(X)$. We shall denote the adjoint of U_T by U_T^* $(U_T^* f(x) = E(f|T^{-1}\mathcal{B})(T^{-1}x))$ and we write $U_{T_n}^* \to U_T^*$ to denote convergence in the strong operator topology. We have $U_{T_n}^* \to U_T^*$ implies $U_{T_n} \to U_T$ (see Proposition 1) but although $U_{T_n} \to U_T$ implies $U_{T_n}^* \to U_T^*$ is valid for automorphisms it is not true for endomorphisms. For this reason we consider the following two topologies on $\mathcal{E}(X)$. (From now on we write Tf for $U_T f$.)

(i) <u>Weak Topology</u>. A neighbourhood of $S \in \mathcal{E}(X)$ is specified by a finite set $f_1,\ldots,f_k$ of members of $L^2(X)$ and some $\varepsilon > 0$: $O(S;f_i,\ldots,f_k;\varepsilon) = \{T \in \mathcal{E}(X) \mid \|Sf_i - Tf_i\| < \varepsilon,\ 1 \le i \le k\}$. A sequence

$\{T_n\}$ converges to S in the weak topology if $\|T_nf - Sf\| \to 0$ for all $f \in L^2(X)$.

(ii) Strong Adjoint Topology. A neighbourhood of $S \in \mathcal{E}(X)$ is specified by a finite set $f_1,\ldots,f_k$ of members of $L^2(X)$ and some $\varepsilon > 0$: $U(S;f_1,\ldots,f_k;\varepsilon) = \{T \in \mathcal{E}(X) \mid \|Sf_i - Tf_i\| < \varepsilon$ and $\|S^*f_i - T^*f_i\| < \varepsilon,\ 1 \le i \le k\}$. A sequence $\{T_n\}$ converges to S in this topology if $\|T_nf - Sf\| \to 0$ and $\|T_n^*f - S^*f\| \to 0$ for all $f \in L^2(X)$.

Proposition 1: Let $\{T_n\}$ be a sequence in $\mathcal{E}(X)$ and let $S \in \mathcal{E}(X)$. Then the statement "T_n converges to S in the strong adjoint topology" is equivalent to each of the following

(a) $\|T_n^* - S^*f\| \to 0$ for all $f \in L^2(X)$

(b) $\|T_nf - Sf\| \to 0$ and $\|E(f|T_n^{-1}\mathcal{B}) - E(f|S^{-1}\mathcal{B})\| \to 0$ for all $f \in L^2(X)$.

Proof: (a) We have to show $\|T_n^*f - S^*f\| \to 0$ for all $f \in L^2(X)$ implies $\|T_nf - Sf\| \to 0$ for all $f \in L^2(X)$. If $f \in L^2(X)$

$$\begin{aligned}\|T_nf - Sf\|^2 &= 2\|f\|^2 - \langle T_nf,Sf\rangle - \langle Sf,T_nf\rangle \\ &= 2\|f\|^2 - \langle f,T_n^*Sf\rangle - \langle T_n^*Sf,f\rangle \\ &\to 2\|f\|^2 - \langle f,S^*Sf\rangle - \langle S^*Sf,f\rangle \\ &= 2\|f\|^2 - \langle Sf,Sf\rangle - \langle Sf,Sf\rangle = 0.\end{aligned}$$

(b) We have $E(f|T^{-1}\mathcal{B}) = TT^*f$ so

$$\begin{aligned}\|E(f|T_n^{-1}\mathcal{B}) - E(f|S^{-1}\mathcal{B})\| &= \|T_nT_n^*f - SS^*f\| \\ &\le \|T_n(T_n^*f - S^*f)\| + \|(T_n - S)S^*f\| \\ &= \|T_n^*f - S^*f\| + \|(T_n-S)S^*f\|\end{aligned}$$

and

$$\|T_n^*f - S^*f\| = \|T_n(T_n^*f - S^*f)\| \le \|T_nT_n^*f - SS^*f\| + \|(S-T_n)S^*f\|$$

$$= \|E(f|T^{-1}\mathcal{B}) - E(f|S^{-1}\mathcal{B})\| + \|(S-T_n)S^*f\|.$$

We remark that it is sufficient to check (a) or (b) for $f = \chi_A$, $A \in \cup_k \hat{\beta}_k$, where $\{\beta_k\}$ is a sequence of finite partitions whose σ-algebras $\hat{\beta}_k$ increase to $\mathcal{B}$ (i.e., $\hat{\beta}_n \nearrow \mathcal{B}$).

Proposition 2: _Let_ $\{f_n\} \subset L^2(X)$ _be a sequence of functions of norm 1 whose linear span is dense in_ $L^2(X)$. _Then_

$$D(S,T) = \sum_{n=1}^{\infty} \frac{1}{2^n} (\|Sf_n - Tf_n\| + \|S^*f_n - T^*f_n\|)$$

is a complete metric on $E(X)$ _compatible with the strong adjoint topology_.

Proof: The proof is the same as that for automorphisms [3].

We have already remarked that the two topologies (i) and (ii) coincide on $A(X)$. However

Proposition 3: $A(X)$ _is dense in_ $E(X)$ _with respect to the weak topology and is closed and nowhere dense with respect to the strong adjoint topology_.

Proof: Let $\{\beta_n\}$ be an increasing sequence of finite partitions whose σ-algebras $\hat{\beta}_n$ increase to $\mathcal{B}$. To show $A(X)$ is dense in the weak topology it suffices to construct, for each $S \in E(X)$, $k \in Z^+$, $\varepsilon > 0$, some $T \in A(X)$ with $\|T\chi_A - S\chi_A\| < \varepsilon$ for all $A \in \beta_k$. But this is immediate: for each $A \in \beta_k$ define an invertible measure-preserving transformation of $S^{-1}A$ onto A, and the combined transformation T is an automorphism $T\chi_A = S\chi_A$.

We shall now show that $A(X)$ is closed in the strong adjoint topology. If $\{T_n\}$ is a sequence of automorphisms converging to $S \in E(X)$ then by Proposition 1 we have $E(f|T_n^{-1}\mathcal{B}) \to E(f|S^{-1}\mathcal{B})$ in $L^2(X)$ for each $f \in L^2(X)$. But $E(f|T_n^{-1}\mathcal{B}) = f$ for each n and therefore $E(f|S^{-1}\mathcal{B}) = f$ for all $f \in L^2(X)$. Hence $S \in A(X)$, and $A(X)$ is closed.

In Section 2 we shall show that the exact Markov endomorphisms are dense in $E(X)$ and therefore $A(X)$ has no interior.

Proposition 4: <u>The strong-mixing endomorphisms form a set of the first category with respect to the weak topology</u>.

<u>Proof</u>: The proof for automorphisms is contained in [3] and goes over for our case on noting that $A(X)$ is dense in $E(X)$ in the weak topology (Proposition 3).

An endomorphism T is said to be <u>exact</u> if $\bigcap_0^\infty T^{-n}\mathcal{B} = N = \{\emptyset, X\}$. Exact endomorphisms are the "opposites" of automorphisms. As we have said, we shall prove that exact Markov endomorphisms are dense in $E(X)$ with respect to the strong adjoint topology. Assuming this for the moment we prove in contrast to Proposition 4:

Theorem 1: <u>The set of exact endomorphisms is a dense G_δ in $E(X)$ with respect to the strong adjoint topology</u>.

<u>Proof</u>: T is exact if and only if $E(f \mid \bigcap_0^\infty T^{-n}\mathcal{B}) = \int f\,d\mu$ for all $f \in L^2(X)$. We have $E(f|T^{-1}\mathcal{B}) = T^n T^{*n} f$. Let $\{f_i\}^\infty$ be dense in $L^2(X)$. For natural numbers r, i, n let

$$U_{r,i,n} = \{T \in E(X) \mid \|T^n T^{*n} f_i - \int f_i d\mu\| < \frac{1}{r}\}.$$

This is an open set in the strong adjoint topology and therefore $\bigcap_r \bigcap_i \bigcap_N \bigcup_{n \geq N} U_{r,i,n}$ is a G_δ. We claim that this G_δ set is precisely the set of exact endomorphisms. By the Martingale theorem we know $\|E(f|T^{-n}\mathcal{B}) - E(f|\bigcap_0^\infty T^{-j}\mathcal{B}) \to 0$ for all $f \in L^2(X)$. Therefore every exact endomorphism belongs to the G_δ set. Conversely, if T belongs to the G_δ set then for each i and each N there exists $n_N \geq N$ with $\|E(f_i|T^{-n_N}\mathcal{B}) - \int f_i\, d\mu\| < \frac{1}{N}$. Therefore $\|E(f_i|T^{-n_N}\mathcal{B}) - \int f_i\, d\mu\|$ $\to 0$ as $N \to \infty$ and we must have $\int f_i\, d\mu = E(f_i \mid \cap_0^\infty T^{-n}\mathcal{B})$. Since $\{f_i\}$ is a dense subset of $L^2(X)$ we have that T is exact.

§2. Exact Markov Endomorphisms are Dense

If A, B are matrices of the same size we will write $|A - B|$ for the maximum absolute value of the entries of $A - B$.

Let P be a stochastic $k \times k$ matrix with strictly positive left fixed vector p, the sum of whose elements equals 1. Such a matrix cannot have trivial rows or columns and will be called non-trivial. It is well known that up to a permutation equivalence P may be written

$$P = \begin{pmatrix} A & & & \\ & B & & \\ & & C & \\ & & & \ddots \end{pmatrix}$$

where $A, B, C, \ldots$ are irreducible and all other entries are zeros. We write, correspondingly, $p = (a,b,c,\ldots)$. For convenience of presentation we shall assume that P is composed of three such irreducible matrices.

Lemma 1: <u>For any $\varepsilon > 0$ there exists an aperiodic irreducible stochastic matrix Q with strictly positive left fixed vector q (whose elements sum to 1) such that</u> $|P - Q| < \varepsilon$, $|p - q| < \varepsilon$.

<u>Proof</u>: The matrix A, being irreducible, has a unique left fixed vector a whose entries sum to $\|a\|$. The compactness of stochastic matrices and vectors with a given sum implies that there exist

strictly positive matrices A' with left fixed vector a' as close as we like to A, a, respectively, such that $\|a'\| = \|a\|$. The same is true, of course, for B and C. To prove the lemma, then, we may assume without loss of generality that A, B, C are strictly positive.

Let $P = \begin{pmatrix} A & & \\ & B & \\ & & C \end{pmatrix}$ be written as

$$\left(\begin{array}{ccc|ccc|ccc}
a_{11} & \cdots & a_{1r} & & & & & & \\
 & & & & & & & & \\
a_{r1} & \cdots & a_{rr} & & & & & & \\
\hline
 & & & b_{11} & \cdots & b_{1s} & & & \\
 & & & & & & & & \\
 & & & b_{s1} & \cdots & b_{ss} & & & \\
\hline
 & & & & & & c_{11} & \cdots & c_{1t} \\
 & & & & & & & & \\
 & & & & & & c_{t1} & \cdots & c_{tt}
\end{array}\right)$$

and let P_ε =

$$\left(\begin{array}{ccc|ccc|ccc}
a_{11}-\rho & \cdots & a_{1r} & \rho & & & & & \\
 & & & & & & & & \\
a_{r1} & \cdots & a_{rr} & & & & & & \\
\hline
 & & & b_{11}-\sigma & \cdots & b_{1s} & \sigma & & \\
 & & & & & & & & \\
 & & & b_{s1} & \cdots & b_{ss} & & & \\
\hline
\tau & & & & & & c_{11}-\tau & \cdots & c_{1t} \\
 & & & & & & & & \\
 & & & & & & c_{t1} & \cdots & c_{tt}
\end{array}\right)$$

The subtractions above occur only in 3 places and are compensated for (to keep P_ε stochastic) in 3 places.

Suppose $a = (a_1,\dots,a_r)$, $b = (b_1,\dots,b_s)$ and $c = (c_1,\dots,c_t)$. Then

$$(a,b,c)P_\varepsilon$$

$$= (a_1(1-\rho)+\tau c_1,\ a_2,\dots,a_r;\ b_1(1-\sigma)+\rho a_1, b_2,\dots,b_s;\ c_1(1-\tau)+\sigma b_1, c_2,\dots,c_t),$$

so $pP_\varepsilon = p$ if

$$a_1(1-\rho)+\tau c_1 = a_1, \quad b_1(1-\sigma)+a_1\rho = b_1, \quad c_1(1-\tau)+b_1\sigma = c_1.$$

Hence $pP_\varepsilon = p$ if $\tau c_1 = a_1\rho$, $a_1\rho = b_1\sigma$, $b_1\sigma = c_1\tau$.

We therefore choose $\tau < \max(\varepsilon, a_1\varepsilon/c_1, b_1\varepsilon/c_1)$ and define $\rho = \dfrac{\tau c_1}{a_1}$, $\sigma = \dfrac{\tau c_1}{b_1}$. Then $|P - P_\varepsilon| < \varepsilon$ and $pP_\varepsilon = p$.

Lemma 2: <u>Let $\{\beta_n\}$ be an increasing sequence of finite partitions such that $\hat{\beta}_n \nearrow \mathcal{B}$. The sets</u>

$$S(k,n,\delta,\varepsilon) = \{T \in E(X) \mid \|E(\chi_A|S^{-1}\hat{\beta}_n) - E(\chi_A|T^{-1}\mathcal{B})\| < \varepsilon \text{ for all } A \in \beta_k \text{ and } \|S\chi_B - T\chi_B\| < \delta \text{ for all } B \in \beta_n\}$$

<u>for $k < n \in Z^+$, $\delta > 0$, $\varepsilon > 0$, form a fundamental system of neighbourhoods of S in the strong adjoint topology.</u>

<u>Proof</u>: From Proposition 1 we know $\{T \mid \|S^*\chi_A - T^*\chi_A\| < \varepsilon$ for all $A \in \beta_k\}$ form a fundamental system of neighbourhoods of S in the strong adjoint topology. Let $k \in Z^+$ and $\varepsilon > 0$ be given.

$$\begin{aligned}\|S^*\chi_A - T^*\chi_A\| &= \|TS^*\chi_A - TT^*\chi_A\| \\ &\le \|TS^*\chi_A - SS^*\chi_A\| + \|SS^*\chi_A - TT^*\chi_A\| \\ &= \|TS^*\chi_A - SS^*\chi_A\| + \|E(\chi_A|S^{-1}\mathcal{B}) - E(\chi_A|T^{-1}\mathcal{B})\| \\ &\le \|TS^*\chi_A - SS^*\chi_A\| + \|E(\chi_A|S^{-1}\hat{\beta}_n) - E(\chi_A|T^{-1}\mathcal{B})\| + \varepsilon/3\end{aligned}$$

if n is large enough.

Choose $\delta > 0$ and choose $n > k$ so that the above inequality holds and also so that $\|Tx_B - Sx_B\| < \delta$ for all $B \in \beta_n$ implies $\|TS^*x_A - SS^*x_A\| < \varepsilon/3$ for all $A \in \hat{\beta}_k$. Then $S(k,n,\delta,\varepsilon/3) \subset \{T \mid \|S^*x_A - T^*x_A\| < \varepsilon\}$.

We shall call $T \in E(X)$ a Markov endomorphism if it is isomorphic to the one-sided shift on a stationary Markov chain with a finite number of states.

Theorem 2: <u>Exact Markov endomorphisms are dense in</u> $E(X)$ <u>with respect to the strong adjoint topology</u>.

Proof: Let $S \in E(X)$ and let $\{\beta_n\}$ be an increasing sequence of finite partitions so that $\hat{\beta}_n \nearrow B$. Fix $\varepsilon > 0$, $\delta > 0$ and $k,n \in Z^+$ with $k < n$. We shall construct an exact Markov endomorphism T in the neighbourhood $S(k,n,\delta,\varepsilon)$ of S (see Lemma 2).

Let $\beta_n = \{B_1,\ldots,B_N\}$ where $m(B_i) > 0$ all i, and define the non-trivial stochastic matrix P by $P(i,j) = \dfrac{m(B_i \cap S^{-1}B_j)}{m(B_i)}$. P has a left fixed probability vector p $(pP = p)$ given by $p(i) = m(B_i)$. Let $\eta > 0$ be chosen later and, using Lemma 1, let $\tilde{P}$ be an irreducible aperiodic stochastic matrix $\tilde{P}$ with left fixed probability vector $\tilde{p}$ such that $|P - \tilde{P}| < \eta$ and $|p - \tilde{p}| < \eta$. Let $\tilde{m}$ denote the shift invariant Markov probability measure defined on $\tilde{X} = \prod_{n=0}^{\infty} \{1,\ldots,N\}$ by $\tilde{P}$, $\tilde{p}$, and let $\tilde{S}$ denote the shift on $\tilde{X}$.

If $[i,j] = \{\tilde{x} \mid \tilde{x}_0 = i, \tilde{x}_1 = j\}$ then $\tilde{m}([i,j]) = \tilde{p}(i)\tilde{P}(ij)$ so that $|m(B_i \cap S^{-1}B_j) - \tilde{m}([i,j])| < 2\eta$. Let $\varphi\colon X \to \tilde{X}$ be an isomorphism which maps $B_i \cap S^{-1}B_j$ into $[i,j]$ if $m(B_i \cap S^{-1}B_j) \le \tilde{m}([i,j])$ and whose inverse maps $[i,j]$ into $B_i \cap S^{-1}B_j$ otherwise. Now put $T = \varphi^{-1}\tilde{S}\varphi$.

We have

$$\left\|x_{B_j} - x_{\varphi^{-1}[j]}\right\| = \left\|\sum_k (x_{B_j \cap S^{-1}B_k} - x_{\varphi^{-1}[j,k]})\right\| \le N(2\eta)^{1/2},$$

and

$$\left\|\chi_{B_j}\circ S - \chi_{B_j}\circ T\right\| = \left\|\sum_i (\chi_{B_i\cap S^{-1}B_j} - \chi_{B_i\cap T^{-1}B_j})\right\|$$

$$\leq \left\|\sum_i (\chi_{B_i\cap S^{-1}B_j} - \chi_{\varphi^{-1}[i,j]})\right\| + \left\|\chi_{\varphi^{-1}[j]}\circ T - \chi_{B_j}\circ T\right\|$$

$$\leq 2N(2\eta)^{1/2} .$$

Therefore

$$\left\|E(\chi_{B_i}|S^{-1}\hat{\beta}_n) - E(\chi_{B_i}|T^{-1}\mathcal{B})\right\|$$

$$\leq \left\|\sum_j \chi_{B_j}\circ S\,\frac{m(B_i\cap S^{-1}B_j)}{m(B_j)} - E(\chi_{\varphi^{-1}[i]}|T^{-1}\mathcal{B})\right\| + \left\|E(\chi_{\varphi^{-1}[i]} - \chi_{B_i})|T^{-1}\mathcal{B})\right\|$$

$$\leq \left\|\sum_j \chi_{B_j}\circ S\,\frac{m(B_i\cap S^{-1}B_j)}{m(B_j)} - \sum_j \chi_{\varphi^{-1}[j]}\circ T\,\frac{\tilde{p}(i)\tilde{P}(i,j)}{\tilde{p}(j)}\right\| + \left\|\chi_{\varphi^{-1}[i]} - \chi_{B_i}\right\|$$

$$\leq \sum_j \left\|\chi_{B_j}\circ S - \chi_{\varphi^{-1}[j]}\circ T\right\| + \sum_j \left|\frac{p(i)P(i,j)}{p(j)} - \frac{\tilde{p}(i)\tilde{P}(i,j)}{\tilde{p}(j)}\right| + \left\|\chi_{\varphi^{-1}[i]} - \chi_{B_i}\right\|$$

$$\leq \sum_j \left\|\chi_{B_j}\circ S - \chi_{B_j}\circ T\right\| + \sum_j \left\|\chi_{B_j}\circ T - \chi_{\varphi^{-1}[j]}\circ T\right\|$$
$$+ \sum_j \left[\frac{1}{p(j)}\left|p(i)P(i,j) - \tilde{p}(i)\tilde{P}(i,j)\right| + \tilde{p}(i)\tilde{P}(i,j)\left|\frac{1}{p(j)} - \frac{1}{\tilde{p}(j)}\right|\right]$$
$$+ N(2\eta)^{1/2}$$

$$\leq 3N^2(2\eta)^{1/2} + N(2\eta)^{1/2} + 2\eta \sum_j \frac{1}{p(j)} + \eta \sum_j \frac{1}{p(j)}[p(j) - \eta].$$

Hence if $A \in \beta_n$ then A is a union of some members of β_n and

$$\|E(\chi_A|S^{-1}\hat{\beta}_n) - E(\chi_A|T^{-1}\mathcal{B})\| \le N(3N^2(2\eta)^{1/2} + N(2\eta)^{1/2} + 2\eta\sum_j \frac{1}{p(j)} + \eta\sum_j \frac{1}{p(j)[p(j)-\eta]}).$$

Now choose $\eta > 0$ so that this latter quantity is less than ε (and $\eta < \min_j m(B_j)$) and so that $2N(2\eta)^{1/2} < \delta$. Then $T \in S(t,n,\delta,\varepsilon)$.

T is exact because one-sided Markov shifts on aperiodic irreducible Markov chains are exact [1].

§3. Irregular Endomorphisms

We say that a $k \times k$ irreducible stochastic matrix P, with invariant initial probability p, is irregular if the function on $\prod_{n=0}^{\infty}\{1,\ldots,k\}$ defined by $x \to p(x_0)P(x_0x_1)/p(x_1)$ generates (up to sets of measure zero) the full σ-algebra. This is the case, of course, if the matrix P', given by $P'(i,j) = p(i)P(i,j)/p(j)$, satisfies $P'(i,j) \ne P'(k,\ell)$ when $(i,j) \ne (k,\ell)$.

Clearly irregularity is a generic property for $k \times k$ matrices. This might suggest that irregularity is generic among all endomorphisms if we define $T \in \mathcal{E}(X)$ to be irregular if $E_T = E(\mathcal{B}|T^{-1}\mathcal{B})$ generates the full σ-algebra. (If α is a finite partition then $E(\alpha|T^{-1}\mathcal{B}) = \sum_{A\in\alpha}\chi_A m(A|T^{-1}\mathcal{B})$ and $E(\mathcal{B}|T^{-1}\mathcal{B}) = \lim_{n\to\infty} E(\beta_n|T^{-1}\mathcal{B})$ for any increasing sequence $\{\beta_n\}$ of finite partitions with $\hat{\beta}_n \nearrow \mathcal{B}$. $I(\mathcal{B}|T^{-1}\mathcal{B}) = -\log E(\mathcal{B}|T^{-1}\mathcal{B})$ is the information function of T [7]. An irregular endomorphism T is characterised up to isomorphism by the invariants

$$\chi_n(s_1,\ldots,s_n) = \int \exp 2\pi i(s_1E_T(x) + \ldots + s_nE_T(T^{n-1}x))\,dm,$$

$$n = 1,2,\ldots,$$

because T is isomorphic to the shift on $\prod_{n=0}^{\infty} R$ endowed with the probability m_T given by $m_T(F) = m(\{x \mid (E_T(x),E_T(Tx),\ldots) \in F\})$. This measure is characterised by the n-fold characteristic functions

x_n, $n \geq 1$.

A slight modification of our proof that exact Markov endomorphisms are dense in $E(X)$ shows that irregular exact Markov endomorphisms are also dense.

However, irregularity is <u>not</u> generic. In fact

Theorem 3. *The set of endomorphisms T with $E_T = 0$ a.e. (or $I_T = I(\mathcal{B}|T^{-1}\mathcal{B}) = \infty$ a.e.) is a dense G_δ, so that irregular endomorphisms form a set of first category.*

Proof: The set of endomorphisms T with $E_T = 0$ a.e. is

$$\bigcap_n \bigcup_{\alpha \text{ finite}} \{T \mid \int E(\alpha|T^{-1}\mathcal{B})\, dm < \frac{1}{n}\} .$$

Since $E(\chi_A|T^{-1}\mathcal{B}) = TT^*\chi_A$, the above set is a G_δ. We have to prove that the above set is dense.

Let S be an exact Markov endomorphism with generator α so that $\alpha^n = \alpha \vee \ldots \vee S^{-n}\alpha$ has the property that $\hat{\alpha}^n \nearrow \mathcal{B}$. We shall find endomorphisms T arbitrarily close to S with the property that $E(\mathcal{B}|T^{-1}\mathcal{B}) = 0$. This will complete the proof by virtue of Theorem 2. It suffices, for arbitrarily large n, to produce such an endomorphism T with the property that $T\chi_A = S\chi_A$ and $E(\chi_A|S^{-1}\mathcal{B}) = E(\chi_A|T^{-1}\mathcal{B})$ for all $A \in \alpha^n$. Since S is Markov with respect to the generator α^n, there is no loss in generality, if we do this for an arbitrary Markov generator, for which we shall retain the symbol α.

Let S_1 be an endomorphism of a Lebesgue space $(X_1,\mathcal{B}_1,m_1)$ with the property that $I(\mathcal{B}_1|S_1^{-1}\mathcal{B}_1) = \infty$ a.e. For example S_1 could be the Bernoulli endomorphism with unit interval as state space. Define $(\tilde{X},\tilde{\mathcal{B}},\tilde{m}) = (X,\mathcal{B},m) \times (X_1,\mathcal{B}_1,m_1)$ and $\tilde{T} = S \times S_1$. Corresponding to the partition $\{A \cap S^{-1}B : A,B \in \alpha\}$ of X we have a partition $\{(A \cap S^{-1}B) \times X_1 : A,B \in \alpha\}$ of $\tilde{X}$. Let φ be an isomorphism of X onto $\tilde{X}$ which maps $A \cap S^{-1}B$ onto $(A \cap S^{-1}B) \times X_1$ when $A,B \in \alpha$. Now we define $Tx = \varphi^{-1}\tilde{T}\varphi(x)$ so that $T^{-1}B = S^{-1}B$ and hence

$\chi_B \circ S = \chi_B \circ T$ for $B \in \alpha$. Finally

$$E(\chi_B | S^{-1}\mathcal{B}) = E(\chi_B | S^{-1}\hat{\alpha})$$

when $B \in \alpha$ and

$$\begin{aligned} E(\chi_B | T^{-1}\mathcal{B}) &= E(\chi_{\varphi(B)} | \varphi T^{-1}\mathcal{B}) \circ \varphi = E(\chi_{B \times X_1} | S^{-1}(\mathcal{B} \times \mathcal{B}_1)) \circ \varphi \\ &= E(\chi_B | S^{-1}\mathcal{B}) \circ \pi_1 \circ \varphi \qquad (\text{where } \pi_1(x, x_1) = x) \\ &= E(\chi_B | S^{-1}\hat{\alpha}) \circ \pi_1 \circ \varphi = E(\chi_B \circ \pi_1 \circ \varphi | \varphi^{-1} \circ \pi_1^{-1} \circ S^{-1}\hat{\alpha}) \\ &= E(\chi_{\varphi^{-1}(B \times X_1)} | \varphi^{-1}(S^{-1}\hat{\alpha} \times X_1)) \\ &= E(\chi_B | S^{-1}\hat{\alpha}) = E(\chi_B | S^{-1}\mathcal{B}). \end{aligned}$$

All that remains is to show that $I(\mathcal{B} | T^{-1}\mathcal{B}) = \infty$ a.e.

But $I(\tilde{\mathcal{B}} | \tilde{T}^{-1}\tilde{\mathcal{B}}) \circ \varphi = I(\mathcal{B} | T^{-1}\mathcal{B})$ and

$$I(\tilde{\mathcal{B}} | \tilde{T}^{-1}\tilde{\mathcal{B}})(x, x_1) = I(\mathcal{B} | S^{-1}\mathcal{B})(x) + I(\mathcal{B}_1 | S_1^{-1}\mathcal{B}_1)(x_1) = \infty \text{ a.e.}$$

The Bernoulli endomorphism which has the unit interval (with Lebesgue measure) as its state space will be called the uniform Bernoulli endomorphism.

Corollary. <u>The set of exact endomorphisms with infinite entropy contains a dense G_δ. The set of exact endomorphisms having the uniform Bernoulli endomorphism as a factor contains a dense G_δ.</u>

<u>Proof</u>: Theorems 1 and 3 show that the set of exact endomorphisms T of X with $I(\mathcal{B} | T^{-1}\mathcal{B}) = \infty$ a.e. form a dense G_δ. Such endomorphisms have the uniform Bernoulli endomorphism as a factor because Rohlin's measure theory [9] gives a non-atomic σ-algebra $\mathcal{C}$ as an independent complement to $T^{-1}\mathcal{B}$ (i.e., is independent of $T^{-1}\mathcal{B}$ and $\mathcal{C} \vee T^{-1}\mathcal{B} = \mathcal{B}$).

It T is an exact endomorphism and $\mathcal{B} = \mathcal{C} \vee T^{-1}\mathcal{B}$ where $\mathcal{C}$ is non-atomic and $\mathcal{C}$ and $T^{-1}\mathcal{B}$ are independent, it is natural to ask if T is isomorphic to the uniform Bernoulli endomorphism. This kind of problem appeared in [12] where the impression is given that the answer is affirmative. However, the answer is negative as the following reasoning shows. Take an exact endomorphism S that does not have a Bernoulli natural extension. (S exists because Ornstein [6] has shown the existence of Kolmogorov atuomorphisms which are not Bernoulli shifts.) Let V denote the uniform Bernoulli endomorphism and put $T = S \times V$. Then $I(\mathcal{B}\, T^{-1}\mathcal{B}) = \infty$ so that a non-atomic $\mathcal{C}$ exists with $\mathcal{C} \vee T^{-1}\mathcal{B} = \mathcal{B}$ and $\mathcal{C}$ independent of $T^{-1}\mathcal{B}$. However T is not isomorphic to V because any factor of the uniform Bernoulli endomorphism has a Bernoulli natural extension [6].

Rosenblatt [10] has discussed related problems when $\mathcal{C}$ is atomic. The answer is negative in this case too. (See also [8].)

§4. g-measures

In this section we consider some problems analogous to those of the earlier sections. Rather than fixing a measure and considering all transformations preserving it, we shall consider a fixed continuous transformation $T: X \to X$ of a compact space and study all the T-invariant Borel probability measures on X whose information function is continuous in a strong sense.

We shall study the case where $T: X \to X$ is a one-sided subshift of finite type. This means there is a finite set C, with $|C|$ elements, and a $|C| \times |C|$ matrix A whose entries are zeros and ones so that X is the subset of C^{Z^+} defined by $x = \{x_n\}_0^\infty \in X$ if and only if $A(x_n, x_{n+1}) = 1$ for all $n \ge 0$. If C is given the discrete topology and C^{Z^+} the product topology then X is a closed subset of the compact metrisable space C^{Z^+}. $T: X \to X$ is defined by $(Tx)_n = x_{n+1}$, $n \ge 0$. T is a local homeomorphism. We shall always assume T

is topologically mixing, which is equivalent to assuming there exists $n > 0$ with A^n having all entries strictly positive.

Let $\mathcal{G} = \{g \in C(X) \mid g > 0$ and $\sum_{y \in T^{-1}x} g(y) = 1, \; \forall x \in X\}$. If we equip $\mathcal{G}$ with the metric ρ defined by $\rho(g_1, g_2) = \|\log g_1 - \log g_2\|_\infty$ ($\|\cdot\|_\infty$ denotes the supremum norm on $C(X)$), then $\mathcal{G}$ is a complete metric space. For $g \in \mathcal{G}$ we can define $L_g: C(X) \to C(X)$ by $L_g f(x) = \sum_{y \in T^{-1}x} g(y) f(y)$. Since L_g is a positive operator and $L_g(1) = 1$, its dual L_g^* maps the compact convex set $M(X)$ of all Borel probabilities on X into itself. Therefore L_g^* always has at least one fixed point in $M(X)$. Any member of $M(X)$ which is a fixed point of L_g^* is called a g-measure, [4]. These measures are important in the study of equilibrium states, [11]. Let $M(T)$ denote those members of $M(X)$ which are T-invariant.

Lemma 3. ([5], [11].)

(i) μ _is a_ g-_measure_ _if and only if_ $\mu \in M(T)$ _and_ $I_\mu(\mathcal{B} | T^{-1}\mathcal{B}) = -\log g$.

(ii) μ _is a_ g-_measure_ _if and only if_ $\mu \in M(T)$ _and_ $h_\mu(T) = -\int \log g \, d\mu$.

(iii) _A_ g-_measure_ _has support_ X.

Let $M_{\mathcal{G}}$ denote the collection of all g-measures as g runs through $\mathcal{G}$. We have a natural map $\pi: M_{\mathcal{G}} \to \mathcal{G}$ given by $\pi(\mu) = g$ if μ is a g-measure. It is unknown if this map π is injective or not. However, it is known that $\pi^{-1}(g)$ is a singleton for g in a dense subset of $\mathcal{G}$, ([4], [11]). We shall show in Theorem 5 that $\{g \in \mathcal{G} \mid \pi^{-1}(g)$ is a singleton$\}$ contains a dense G_δ. If $\mathcal{G}_k$ denotes the set of those g depending only on the first k coordinates (i.e., k is the least natural number with the property $g(x) = g(y)$ if $x_i = y_i$, $0 \le i \le k-1$) then $\pi^{-1}(g)$ is a singleton for $g \in \mathcal{G}_k$. $\cup_k \mathcal{G}_k$

is dense in $\mathcal{G}$. If μ is a g-measure for $g \in \mathcal{G}_k$ then μ is a $k-1$ step Markov measure. We shall call the members of this family $(\pi^{-1}(\cup_k \mathcal{G}_k))$ <u>strong Markov measures</u>. They are supported on X.

Let d denote a metric on $M(T)$ giving the weak*-topology. $M(T)$ is compact with this topology. Define a metric D on M_g by $D(\mu,\nu) = d(\mu,\nu) + \rho(\pi(\mu),\pi(\nu))$.

Lemma 4. M_g <u>is complete with respect to the metric</u> D.

<u>Proof</u>: Suppose $\{\mu_n\}$ is a Cauchy sequence for D. Then $\{\mu_n\}$ is Cauchy for d and hence $\mu_n \longrightarrow \mu \in M(T)$. Also $\|\log g_n - \log g\|_\infty \to 0$ for some $g \in \mathcal{G}$. It remains to show μ is a g-measure and this follows because

$$\int L_g h \, d\mu = \lim_{n\to\infty} \int L_{g_n} h \, d\mu_n = \lim_{n\to\infty} \int h \, d\mu_n = \int h \, d\mu \quad ,$$

for all $h \in C(X)$.

The map $\pi: M_g \to \mathcal{G}$ is clearly continuous. The topology on M_g given by D is strictly stronger than that given by d. We shall illustrate this when $X = \{0,1\}^{Z^+}$. Define g_n by

$$g_n(x) = \begin{cases} 3/4 & \text{if } (x_0,\ldots,x_{n-1}) = (0,0,0,\ldots,0) \\ 1/4 & \text{if } (x_0,\ldots,x_{n-1}) = (1,0,0,\ldots,0) \\ 1/2 & \text{otherwise.} \end{cases}$$

If μ_n is the unique g_n-measure then μ_n is a (n-1)-step Markov measure and one can show that $\mu_n \xrightarrow{d} \mu$ where μ is the product measure with weights (1/2, 1/2). μ is the g-measure for $g = 1/2$ but $\|\log g_n - \log 1/2\|_\infty \not\to 0$.

We shall denote the cylinder set $\{x \mid x_i = a_i, \ 0 \le i \le n-1\}$ by $[a_0,a_1,\ldots,a_{n-1}]$.

Proposition 5. Let μ be a g-measure and define $g_n \in \mathcal{G}$ by

$$g_n(x) = \frac{\mu([x_0,x_1,\ldots,x_{n-1}])}{\mu([x_1,\ldots,x_{n-1}])}.$$

Then $\|\log g_n - \log g\|_\infty \to 0$.

Proof: Since g is uniformly continuous we have

$$c_n = \sup\{\frac{g(w)}{g(z)} \mid w_i = z_i, \quad 0 \le i \le n-1\} \to 1 \qquad \text{as } n \to \infty.$$

Since $\frac{d\mu T}{d\mu} = \frac{1}{g}$ we have

$$\mu([x_1,\ldots,x_{n-1}]) = \int_{[x_0,\ldots,x_{n-1}]} \frac{1}{g(z)}\, d\mu(z)$$

and therefore

$$\frac{g(x)}{c_n} \le \frac{\mu([x_0,\ldots,x_{n-1}])}{\mu([x_1,\ldots,x_{n-1}])} \le c_n g(x).$$

This gives $\|g_n - g\|_\infty \to 0$ and $\|\log g_n - \log g\|_\infty \to 0$.

Proposition 6. The strong Markov measures are dense in M_g (using the metric D).

Proof: Let μ be a g-measure and define g_n as above. Let μ_n be the unique g_n-measure. It remains to show $\mu_n \xrightarrow{d} \mu$. It suffices to show that for any given cylinder $[a_0,\ldots,a_k]$, $\mu_n([a_0,\ldots,a_k]) \to \mu([a_0,\ldots,a_k])$. This follows because if $n > k$ we have $\mu_n([a_0,\ldots,a_k]) = \mu([a_0,\ldots,a_k])$.

Let us call g irregular if $\{g(T^n)\}_{n=0}^\infty$ separates points of X and call $\mu \in M_g$ irregular if $\pi(\mu)$ is irregular. It is clear that if $g \in \mathcal{G}_k$ and g takes distinct values on the cylinders of length k then g is irregular.

Proposition 7. The irregular strong Markov measures are dense in M_g (using the metric D).

Proof: If μ is a g-measure for $g \in \mathcal{g}_k$ then approximate g by some $g' \in \mathcal{g}_k$ with distinct values on k-cylinders. The unique g'-measure μ' will be close to μ in the D metric.

Theorem 4. The subset of M_g consisting of irregular measures contains an open dense set (using the metric D).

Proof: Let $g_0 \in \mathcal{g}_k$ take distinct values on k-cylinders. We shall show there is a neighbourhood U of g_0 in $\mathcal{g}$ which consists of irregular g's. Then $\pi^{-1}(U)$ is open in M_g and the proof will be complete.

Let $\varepsilon = \min\{|\log g_0([x_0,\ldots,x_{k-1}]) - \log g_0([z_0,\ldots,z_{k-1}])| \mid [x_0,\ldots,x_{k-1}] \neq [z_0,\ldots,z_{k-1}]\} > 0$. Let $U = \{g \in \mathcal{g} \mid \rho(g,g_0) < \frac{\varepsilon}{4}\}$. Let $g \in U_0$. If $x \neq z$ there is some j with $g_0(T^j x) \neq g_0(T^j z)$ and so $|\log g(T^j x) - \log g(T^j z)| \geq |\log g_0(T^j x) - \log g_0(T^j z)| - \frac{\varepsilon}{2} \geq \frac{\varepsilon}{2}$. Therefore each $g \in U$ is irregular.

If μ is an irregular g-measure then the map $x \xrightarrow{\varphi_g} (g(x),g(Tx),\ldots)$ is a homeomorphism of X onto a closed subset of $\prod_{n=0}^{\infty} [0,1]$ which conjugates T with the shift on $\prod_0^{\infty} [0,1]$. The map φ_g takes μ to a measure on $\prod_0^{\infty} [0,1]$ which is characterised by its n-fold characteristic functions (see Section 3). Therefore an open dense set of g-measures are characterised by a countable number of invariants. If μ' is an irregular g'-measure and μ, μ' have the same invariants then $\varphi_{g'}^{-1} \circ \varphi_g : X \to X$ is a homeomorphism mapping μ to μ' and commuting with T. Therefore, any two irregular members of M_g with the same invariants are related by a homeomorphism of X commuting with T.

We next show that 'most' $g \in \mathcal{g}$ have a unique g-measure and that 'most' $\mu \in M_g$ are exact.

Theorem 5. $\{g \in \mathcal{g} \mid$ there is a unique g-measure$\}$ contains a dense G_δ in $\mathcal{g}$.

<u>Proof</u>: Let $\{f_n\}_1^\infty$ be dense in $C(X)$. For natural numbers n, m, N and $c \in R$ let $U_{n,m,c,N} = \{g \in \mathcal{G} \mid \|L_g^N f_n - c\|_\infty < \frac{1}{m}\}$. This is an open subset of $\mathcal{G}$ and therefore $\widetilde{\mathcal{G}} = \cap_n \cap_m \cup_c \cup_N U_{n,m,c,N}$ is a G_δ. We claim that

$$\widetilde{\mathcal{G}} = \{g \in \mathcal{G} \mid \text{for all } f \in C(X) \text{ there exists } c(f) \in R \text{ with } \|L_g^i f - c(f)\|_\infty \to 0\}.$$

If g belongs to this set then $g \in \widetilde{\mathcal{G}}$. Conversely, if $g \in \widetilde{\mathcal{G}}$ then for all n, m there exists $c_m(n)$ and there exists N such that $\|L_g^N f_n - c_m(n)\|_\infty < \frac{1}{m}$. Since $\|L_g\|_\infty \le 1$ we have $\|L_g^i f_n - c_m(n)\|_\infty < \frac{1}{m}$ for all $i \ge N$. If μ is any g-measure then $\left|\int f_n \, d\mu - c_m(n)\right| < \frac{1}{m}$ so $\|L_g^i f_n - \int f_n \, d\mu\|_\infty < \frac{2}{m}$ for all $i > N$, and for each n $\|L_g^i f_n - \int f_n \, d\mu\|_\infty \to 0$ as $i \to \infty$. Therefore $\|L_g^i f - \int f \, d\mu\|_\infty \to 0$, for all $f \in C(X)$. Each $g \in \widetilde{\mathcal{G}}$ has a unique g-measure because if $\|L_g^i f - c(f)\|_\infty \to 0$ then $\int f \, d\mu = c(f)$ for each g-measure μ. Since $\cup_{k=1}^\infty \mathcal{G}_k \subset \widetilde{\mathcal{G}}$ ([11]) we know that $\widetilde{\mathcal{G}}$ is dense in $\mathcal{G}$.

If $\mu \in M(T)$ is such that T is an exact endomorphism relative to μ then we will say μ is <u>exact</u>.

Theorem 6. <u>$\{\mu \in M_g \mid \mu$ is exact$\}$ contains a dense G_δ in M_g (with respect to D).</u>

<u>Proof</u>: Let $\{f_n\}_{n=1}^\infty$ be dense in $C(X)$. For natural numbers n, m, N let $U_{n,m,N} = \{\mu \in M_g \mid \|L_{\pi(\mu)}^N f_n - \int f_n \, d\mu\|_\infty < \frac{1}{m}\}$. $U_{n,m,N}$ is an open subset of M_g and therefore $V = \cap_n \cap_m \cup_N U_{n,m,N}$ is a G_δ. Using the fact that $\|L_{\pi(\mu)}\|_\infty \le 1$ we have

$V = \{\mu \in M_g \mid \|L_{\pi(\mu)}^i f - \int f \, d\mu\|_\infty \to 0\}$ for all $f \in C(X)\}$. We claim that each $\mu \in V$ is exact. If $\mu \in V$ then $\int |L^n_{\pi(\mu)}f - \int f\,d\mu|\,d\mu \to 0$ for all $f \in L^1(\mu)$ and therefore $\int |E_\mu(f \mid \cap_0^\infty T^{-n}\mathcal{B}) - \int f\,d\mu|\,d\mu = \lim_{n\to\infty} \int E_\mu(f|T^{-n}\mathcal{B}) - \int f\,d\mu|\,d\mu = \lim_{n\to\infty} \int |U^n_T L^n_{\pi(\mu)}f - \int f\,d\mu|\,d\mu = \lim_{n\to\infty} \int |L^n_{\pi(\mu)}f - \int f\,d\mu|\,d\mu = 0$, for all $f \in L^1(\mu)$. This shows that $\cap_0^\infty T^{-n}\mathcal{B}$ is trivial relative to μ and hence that μ is exact. We know the strong Markov measures are in V and these are dense by Proposition 6.

REFERENCES

1. D. Blackwell and D.A. Freedman, The tail σ-field of a Markov chain and a theory of Orey, Annals. Math. Stat. 35(1964), 1291-1295.

2. J. Feldman, Borel structures and invariants for measurable transformations, Proc. Amer. Math. Soc. 46(1974), 383-394.

3. P. Halmos, Ergodic Theory, Chelsea, 1956.

4. M. Keane, Strongly mixing g-measures, Invent. Math. 16(1972), 309-324.

5. F. Ledrappier, Principe Variationnel et systemes symboliques, Comm. Math. Phys. 33(1973), 119-128.

6. D.S. Ornstein, Ergodic Theory, Randomness and Dynamical Systems, Yale Univ. Press, 1974.

7. W. Parry, Entropy and Generators in Ergodic Theory, Benjamin, 1969.

8. W. Parry and P. Walters, Endomorphisms of a Lebesgue space, Bull. Amer. Math. Soc. 78(1972), 272-276.

9. V.A. Rohlin, On the fundamental ideas of measure theory, Mat. Sborn. 25(1949), 107-150. Amer. Math. Soc. Transl. 71(1952), 1-54.

10. M. Rosenblatt, Stationary processes as shifts of functions of independent random variables, J. Math. and Mech. 8(1959), 665-682.

11. P. Walters, Ruelle's operator theorm and g-measures, Trans. Amer. Math. Soc. 214(1975), 375-387.

12. N. Wiener, Nonlinear Problems in Random Theory, MIT Press, 1958.

UNIVERSITY OF WARWICK

A LINEARIZATION PROCESS FOR FLOWS

by

William Perrizo

This note is concerned with isolating linearity in smooth vector field flows on smooth manifolds.

In [4], a class of smooth flows which display a certain linearity was defined and studied. The linearity condition was formulated in terms of the existence of invariant metric connection forms on the linear frame bundle. A second condition, that the images of these forms be conjugate one-to-another, was added. This condition facilitated the subsequent analysis, but was rather severe. In this paper only a linearity condition is assumed. It is formulated in terms of the existence of a certain matrix valued function on the manifold itself (see 1.6).

In Section 1 a linearization process is developed which can be viewed as a truncation of the nonlinearity of the system. It is shown that a truncatable vector field induces a decomposition of the manifold into regions on which the first variation is homogeneous. That is, a homogeneous region is one on which there is a vector bundle splitting of the tangent bundle into stable, unstable and center subbundles of constant dimensions.

In Section 2 a system of local coordinates is defined which display the stable, unstable and center directions under the flow. It is this system of local coordinates which is used in [4] to study certain topological aspects of the flow.

§1.

All manifolds, maps and vector fields will be considered C^∞ unless stated otherwise. On an n-dimensional Riemannian manifold M; T_xM, L_xM, TM and LM are the tangent space at x, the space of frames at x (ordered bases of T_xM), the manifold of all tangent vectors on M, and the manifold of all frames on M, respectively. The projection maps of tangent vectors or frames to base points will be denoted by π. For a vector space V, Gl(V), gl(V) are the group of nonsingular linear transformations on V, the algebra of all linear transformations on V, respectively. R^n is Euclidean n-space. We will not distinguish between elements of $gl(R^n)$ and their matrices with respect to the canonical basis on R^n. For a map f from one manifold M to another N; df, df(x), and df(x)[v] are the space derivative, the restriction of the space derivative to T_xM, and the image of the tangent vector v under the space derivative, respectively. X[f] is the image of the vector field X operating on the real-valued function f. For a curve c: I → M (I is an interval in R^1), d/dt(c(t)) or c'(t) will denote the velocity vector, dc(t)[1]. The symbols u and w will be used to denote frames or frame fields (written as column vectors). To avoid confusion, the image of x in M under a frame field u will be written u_x since x is the base point of the frame u(x). The symbols $\underline{a}$, $\underline{b}$ will be denote points in R^n (row vectors) or functions from a set U into R^n. The symbols $\overline{a}$, $\overline{b}$ will denote matrices in $Gl(R^n)$ or functions from U to $Gl(R^n)$. The symbols a, b will be used when it doesn't matter. Formal matrix multiplication will be denoted by juxtaposition. A frame u in L_xM induces isomorphisms $\underline{a} \to \underline{a}u$: $R^n \to T_xM$ and $\overline{a} \to \overline{a}u$: $Gl(R^n) \to L_xM$. A frame field u on U induces bijections $\underline{a} \to \underline{a}u$: ($R^n$-valued functions on U) → (vector fields on U) and $\overline{a} \to \overline{a}u$: ($Gl(R^n)$-valued functions on U) → (frame fields on U). We will use u to denote these maps as well as the frame or frame field. Thus, $u^{-1}(Y) = \underline{a}$ if and only if $Y = \underline{a}u$ for example. The frame field u on U combines with the standard frame field e on R^n to form a frame field on $R^n \times U$, which induces the frame field $\overline{\overset{e}{u}}$ on TU via the diffeomorphism $(\underline{a},x) \to \underline{a}u_x\colon R^n \times U \to TU$. Similarly, u combines with the standard frame field on $Gl(R^n)$ (as an open subset of R^{n^2}) to form a frame field on $Gl(R^n) \times U$, which induces the frame field $\overline{\overset{e}{u}}$ on LM via the diffeomorphism $(\overline{a},x) \to \overline{a}u_x\colon Gl(R^n) \times U \to LM$. This diffeomorphism will be denoted by θ_u. We will use $\overset{e}{u}$ for either $\underline{\overset{e}{u}}$ or $\overline{\overset{e}{u}}$ when it is clear from the context or it doesn't matter. Thus, a curve c in TU or LU has velocity

$$c'(t) \;=\; ((u^{-1} \circ c)'(t),\, u^{-1}((\pi c)'(t)))\overset{e}{u}_{c(t)}. \tag{1.0}$$

<u>Definition 1.1</u>. Given a complete vector field X on M and the corresponding flow $[X_t\colon M \to M \mid t \text{ in } R]$, a <u>linear lift</u> of X to LM is a vector field $\overline{H}$ on LM such that $d\pi\,\overline{H} = X$ and $\overline{H}_t(aw) = a\overline{H}_t(w)$ (w in LM, a in $Gl(R^n)$). A linear lift of X to TM is a vector field $\underline{H}$ on TM such that $d\pi\,\underline{H} = X$ and $\underline{H}_t((r\underline{a}+s\underline{b})u) = r\underline{H}_t(\underline{a}u) + s\underline{H}_t(\underline{b}u)$ (r, s in R, u in LM, $\underline{a}$, $\underline{b}$ in R^n).

Clearly, a linear lift $\overline{H}$ to LM induces a linear lift $\underline{H}$ to TM by $\underline{H}_t(\underline{a}w) = \underline{a}\overline{H}_t(w)$ and a linear lift $\underline{H}$ to TM induces a linear lift $\overline{H}$ to LM by

$$\overline{H}_t(w) \;=\; \begin{bmatrix} \underline{H}_t(w^1) \\ \vdots \\ \underline{H}_t(w^n) \end{bmatrix} \qquad (w \text{ in } LM).$$

We will use H for $\underline{H}$ or $\overline{H}$ when the context makes it clear or when it doesn't matter.

<u>Definition 1.2</u>. Let u be a frame on an open set U and let $\overline{H}$ be a linear lift of X. Define ${}^{u}h\colon U \times R^1 \to Gl(R^n)$ by

${}^{u}h(x,t) = u^{-1}(\overline{H}_t(u_x))$ for all t such that $\overline{H}_t(u_x)$ is in TU. Define ${}^{u}H\colon LM \to Gl(R^n)$ by $({}^{u}H(x), u^{-1}(X_x)) = \overset{e}{u}^{-1}(\overline{H}(u_x))$.

We will write $X_t(x)$ as simply xt, then $\overline{H}_t(u_x) = {}^{u}h(x,t)u_{xt}$ and $\overline{H}(u_x) = ({}^{u}H(x), u^{-1}(X_x))\overset{e}{u}_{u_x}$.

Lemma 1.3.

i. <u>For every</u> au_x <u>in</u> LU, $\overline{H}(\bar{a}u_x) = (\bar{a}{}^{u}H(x), u^{-1}(X_x))\overset{e}{\bar{u}}_{\bar{a}u_x}$.

ii. $d/dt({}^{u}h(x,t)) = {}^{u}h(x,t)H(u_{xt})$ <u>where defined</u>.

iii. <u>If</u> w <u>is any other frame field on</u> U <u>then</u> ${}^{w}H$ <u>and</u> ${}^{u}H$ <u>are kinematically similar</u>, $\bar{a}\,{}^{u}H - {}^{w}H\bar{a} = X[\bar{a}]$, <u>where</u> $w = \bar{a}u$.

<u>Proof</u>: Let $c(t) = H_t(au_x) = aH_t(u_x) = a\,{}^{u}h(x,t)u_{xt}$, then

$$H(au_x) = c'(t) = (d/dt(a\,{}^{u}h(x,t))\big|_{t=0}, u^{-1}(X_x))\overset{e}{u}_{au_x}$$

$$= (a\frac{d}{dt}\,{}^{u}h(x,t)\big|_{t=0}, u^{-1}X_x)\overset{e}{u}_{au_x}.$$

Similarly, $H(u_x) = (\frac{d}{dt}\,{}^{u}h(x,t)\big|_{t=0}, u^{-1}X_x)\overset{e}{u}_{u_x}$. Thus, ${}^{u}H(x) = \frac{d}{dt}\,{}^{u}h(x,t)\big|_{t=0}$ and the result i follows.

To prove ii we let $c(t) = H_t(u_x)$ then $c'(t) = H(c(t)) = H({}^{u}h(x,t)u_{xt}) = ({}^{u}h(x,t){}^{u}H(xt), u^{-1}X_{xt})\overset{e}{u}_{c(t)}$. On the other hand, $c'(t) = (\frac{d}{dt}({}^{u}h(x,t)), u^{-1}X_{xt})\overset{e}{u}_{c(t)}$ by (1.0). The result ii follows.

The kinematic similarity of ${}^{u}H$ and ${}^{w}H$ follows from this equation and the general theory of change of basis for linear systems (see [3]).

Lemma 1.4. <u>If</u> $h\colon M \times R \to R$ <u>is such that</u> $h(x,0)$ <u>is independent of</u> x <u>and if</u> $c\colon I \to M$ <u>is a curve</u>,

$$d/dt(h(c(t),t))\big|_{t=0} = d/dt(h(c(0),t))\big|_{t=0}.$$

The proof of this lemma can be found in [1], page 15.

Lemma 1.5. Let H and K be linear lifts of X. The following are equivalent.

i. $[H,K] = 0$.

ii. $K_s \circ H_t = H_t \circ K_s$ (s,t in R).

iii. $K_t \circ H_t = (K+H)_t$ (t in R).

iv. For each frame field u, $X[{}^uK - {}^uH] - [{}^uK, {}^uH] = 0$ $([\bar{a},\bar{b}] = \bar{a}\bar{b} - \bar{b}\bar{a})$.

Proof: The equivalence of i, ii, and iii is proved in most texts treating vector fields and flows ([2], for instance).

Let $\overset{e}{u}{}^i_j$ represent the image under $d\theta_u$ of e_{ij}, the matrix with 1 in row-i column-j and 0 elsewhere. Let $\overset{e}{u}{}^k$ represent the image under $d\theta_u$ of the vector field u^k. Thus, the $\overset{e}{u}{}^i_j$ and $\overset{e}{u}{}^k$ are the $n^2 + n$ components of $\overset{e}{u}$.

To show the equivalence of i and iv, let u be a coordinate frame field on U $(u_i \circ u_j = u_j \circ u_i)$. Then $\overset{e}{u}$ is also a coordinate frame field. Let $u^{-1}X = a$. Since $H(bu_x) = (b^uH(x), u^{-1}X)\overset{e}{u}_{u_x}$,

$$H = \sum_{(i,j)} (u^{-1} \times {}^uH)^i_j \, \overset{e}{u}{}^i_j + \sum_{(k)} a_k \overset{e}{u}{}^k$$

and

$$K = \sum_{(m,p)} (u^{-1} \times {}^uK)^m_p \, \overset{e}{u}{}^m_p + \sum_{(h)} a_h \overset{e}{u}{}^h$$

$(u^{-1} \times {}^uH$ takes bu_x to $b^uH(x))$. For $f\colon LU \to R$,

$$\begin{aligned} H \circ K(f) &= H[\sum_{(m,p)} (u^{-1} \times {}^uK)^m_p \, \overset{e}{u}{}^m_p [f] + \sum_{(h)} a_h \overset{e}{u}{}^h [f] \\ &= \sum_{(m,p)} H[(u^{-1} \times {}^uK)^m_p] \, \overset{e}{u}{}^m_p [f] + \sum_{(m,p)} (u^{-1} \times {}^uK)^m_p \, H[\overset{e}{u}{}^m_p [f]] \\ &\quad + \sum_{(h)} H[a_n] \, \overset{e}{u}{}^h [f] + \sum_{(h)} a_h H[\overset{e}{u}{}^h [f]] \end{aligned}$$

$$= \sum_{(ijmp)} (u^{-1}\times {}^{u}H)^i_j \, \underset{u}{e}{}^{i}_{j} \, [(u^{-1}\times {}^{u}K)^m_p] \, \underset{u}{e}{}^{m}_{p} \, [f] \qquad (1)$$

$$+ \sum_{(kmp)} a_k \, \underset{u}{e}{}^{k} \, [(u^{-1}\times {}^{u}K)^m_p \, \underset{u}{e}{}^{m}_{p} \, [f]$$

$$+ \sum_{(ijmp)} (u^{-1}\times {}^{u}H)^i_j \, (u^{-1}\times {}^{u}K)^m_p \, \underset{u}{e}{}^{i} \circ \underset{u}{e}{}^{m}_{p} [f] \qquad (2)$$

$$+ \sum_{(kmp)} a_k (u^{-1}\times {}^{u}K)^m_p \, \underset{u}{e}{}^{k} \circ \underset{u}{e}{}^{m}_{p} [f] \qquad (3)$$

$$+ \sum_{(ijh)} (u^{-1}\times {}^{u}H)^i_j \, \underset{u}{e}{}^{i}_{j} \, [a_h] \, \underset{u}{e}{}^{h} \, [f] \qquad (4)$$

$$+ \sum_{(hk)} a_k \, \underset{u}{e}{}^{k} \, [a_h] \, \underset{u}{e}{}^{h} \, [f] \qquad (5)$$

$$+ \sum_{(hij)} a_h (u^{-1}\times {}^{u}H)^i_j \, \underset{u}{e}{}^{i}_{j} \circ \underset{u}{e}{}^{h} [f] + \sum_{(kh)} a_k a_h \, \underset{u}{e}{}^{k} \circ \underset{u}{e}{}^{h} [f]. \qquad (6)$$

In $[H,K][f] = H\circ K[f] - K\circ H[f]$ the terms (3), (5), and (6) cancel. Each term in (4) is 0 since a_h is constant on fibers. The expressions (2) take the form

$$\sum_{(ijmp)} ((u^{-1}\times {}^{u}H)^i_j (u^{-1}\times {}^{u}K)^m_p - (u^{-1}\times {}^{u}K)^i_j (u^{-1}\times {}^{u}H)^m_p) \, \underset{u}{e}{}^{i}_{j} \circ \underset{u}{e}{}^{m}_{p} [f]$$

and, thus, each $ijmp$ term cancels with the corresponding $mpij$ term. In the remaining expressions (1), $\underset{u}{e}{}^{i}_{j} [(u^{-1}\times {}^{u}K)] = \underset{u}{e}{}^{i}_{j} [u^{-1}]{}^{u}K$, since ${}^{u}K$ is constant on fibers. Since

$$\begin{aligned} \underset{u}{e}{}^{i}_{j} [u^{-1}](u_x) &= \frac{d}{dt} (u^{-1}((\underset{u}{e}{}^{i}_{j})_t(u_x)))\Big|_{t=0} \\ &= \frac{d}{dt} (u^{-1}((I+te_{ij})u_x))\Big|_{t=0} \\ &= \frac{d}{dt} (I+te_{ij})\Big|_{t=0} = e_{ij}, \end{aligned}$$

we have

$$\underset{u}{e}{}^{i}_{j} [(u^{-1}\times {}^{u}K)^m_p] = (e_{ij}{}^{u}K)^m_p = (\textstyle\sum_{(h)} {}^{u}K^i_h e_{ih})^m_p = \begin{Bmatrix} 0 & \text{if } m \neq i \\ {}^{u}K^i_p & \text{if } m = i \end{Bmatrix}.$$

Also, $\overset{e}{u}^k[u^{-1}\times({}^uK - {}^uH))^m_p] = (u^{-1}\times \overset{e}{u}^k[({}^uK - {}^uH)^m_p]$, since u^{-1} is constant on orbits of $\overset{e}{u}^k$. Thus, from the expressions (1)

$$[H,K][f] = \sum_{(mp)} \Big(\sum_{(j)} (u^{-1}\times {}^uH)^m_j\, {}^uK^j_p - \sum_{(j)} (u^{-1}\times {}^uK)^m_j\, {}^uH^j_p$$

$$+ \sum_{(k)} a_k(u^{-1}\times \overset{e}{u}^k[({}^uK - {}^uH)^m_p])\Big)\, \overset{e}{u}^m_p [f]$$

$$= \sum_{(mp)} \Big((u^{-1}\times {}^uH\,{}^uK)^m_p - (u^{-1}\times {}^uK\,{}^uH)^m_p$$

$$+ u^{-1}\times \sum_{(k)} a_k\, \overset{e}{u}^k[({}^uK - {}^uH)^m_p]\Big)\, \overset{e}{u}^m_p [f]$$

$$= \sum_{(mp)} (u^{-1}\times (X[{}^uK - {}^uH] - [{}^uK, {}^uH])^m_p)\, \overset{e}{u}^m_p [f]$$

$$= (0,\ u^{-1}\times (X[{}^uK - {}^uH] - [{}^uK, {}^uH]))\, \overset{e}{u} [f].$$

If $w = bu$ is an arbitrary frame field on U, ${}^wH = b\,{}^uHb^{-1} - X[b]b^{-1}$ and ${}^wK = b\,{}^uKb^{-1} - X[b]b^{-1}$. Using these identities and straightforward calculation we see that

$$X[{}^wK - {}^wH] - [{}^wK, {}^wH] = b(X[{}^uK - {}^uH] - [{}^uK, {}^uH])b^{-1}.$$

Thus, $[H,K] = 0$ if and only if $X[{}^wK - {}^wH] - [{}^wK, {}^wH] = 0$, for any frame field w on U. The result follows.

<u>Definition 1.6</u>. Let H and K be linear lifts of X. H is a r-truncation of K if:

i. $[H,K] = 0$

ii. About each x in M there is an orthonormal frame field u such that

$${}^uH^i_j = 0 \quad (i < j) \qquad \text{and} \qquad |{}^uH^i_i| \le r \quad (i = 1,\dots,n).$$

It is a simple matter to check that $[\bar{H},\bar{K}] = 0$ if and only if

$[\underline{H},\underline{K}] = 0$.

Theorem 1.7. Let H be an r-truncation of K. To each Y in TM there corresponds a positive number $q(Y)$ such that

$$\|\underline{H}_t Y\| \geq q(Y)\exp(-r|t|) \qquad (t \text{ in } R).$$

Proof: Let $u\colon U \to LU$ be an orthonormal frame field at $x = \pi Y$ as in 1.6 ii. Since ${}^uH(xt)$ has all zeros above the diagonal and $\frac{d}{dt}({}^uh(x,t))\big|_{t=0} = {}^uh(x,t)\,{}^uH(xt)$, $({}^uh(x,t))^i_j = 0$ for $i < j$. Since $|{}^uH^i_j| \leq r$ and $\frac{d}{dt}({}^uh(x,t)^i_i) = {}^uh(x,t)^i_i\,{}^uH(xt)^i_i$, $e^{-r|t|} \leq {}^uh(x,t)^i_i \leq e^{r|t|}$ (note ${}^uh^i_i$ is nonzero since ${}^uh$ is nonsingular). This gives the desired result for small t. This process is then continued along the orbit of x. Clearly, if $t_0 = \sup[t \mid$ the process can be continued to $t] < \infty$, then the above construction applied at xt_0 provides a contradiction. The same argument applies in the negative direction.

Lemma 1.8. Let H be a r-truncation of K.

i. $V = K - H$ is a linear lift of the zero vector field on M.

ii. $V_t(bu_x) = (b\exp(t\,{}^uV(x)))u_x$ (that is, ${}^uv(x,t) = \exp(t\,{}^uVx)$).

iii. If $w = au$ is any other frame field of U, then ${}^wV = a\,{}^uVa^{-1}$.

Proof: The statement i is trivial. The statement iii follows from i and 1.3 iii. To prove ii, we note that, since $u^{-1}\circ V_t \circ u$ is a one-parameter subgroup of $GL(R^n)$, there exists A in $gl(R^n)$ such that $(u^{-1}\circ V_t\circ u)(b) = b(\exp tA)$. Applying u to both sides we have $V_t(bu_x) = b(\exp tA)u_x$. Thus, ${}^uv(x,t) = \exp tA$ and
$A = \frac{d}{dt}({}^uv(x,t))\big|_{t=0} = {}^uv(x,0)\,{}^uV(x) = {}^uV(x)$.

Definition 1.9. Let H, K, V be as above.

i. The eigenvalues of ${}^uV(x)$ with real parts $> r$, $< r$, and in absolute value $\leq r$ will be called r^+, r^-, and r^0 eigenvalues, respectively (by 1.8 iii u can be arbitrary).

ii. $Z^{(\ )}$ will count the $r^{(\ )}$ eigenvalues according to multiplicity, $(\ ) = 0, -, +$.

iii. $W_m = [x \text{ in } M \mid Z^0(x) < m + 1/2]$, $m = 0,\ldots,n$.

iv. A <u>homogeneous region</u> under X is a path component of $W_m - \mathrm{Cl}(W_{m-1})$.

Theorem 1.10.

i. Z^0 <u>is upper semi-continuous</u>.

ii. <u>Each</u> W_m <u>is open and</u> X-<u>invariant</u> $(X_t(W_m) \subseteq W_m)$.

iii. <u>On each homogeneous region, the functions</u> Z^0, Z^+, <u>and</u> Z^- <u>are constant</u>.

<u>Proof</u>: The eigenvalues of ${}^uV$ vary continuously in x. If x is such that $Z^0(x) < m+1/2$, there are at least $n-m-1/2$ non-r^0 eigenvalues for ${}^uV$. This situation persists in a neighborhood of x. Thus, Z^0 is upper semi-continuous and W_m is open.

Since $[H,K] = 0$ and $V = K-H$, it follows that $[V,K] = 0$ and from 1.7 ii that $K_t \circ V_s = V_s \circ K_t$ $(s,t$ in $R)$.

$$\begin{aligned}\frac{d}{ds}(V_s \circ K_t({}^uk(x,t)^{-1}u_x)\Big|_{s=0} &= V(K_t({}^uk(x,t)^{-1}u_x)) = V({}^uk(x,t)^{-1}K_t(u_x)) \\ &= V({}^uk(x,t)^{-1}\ {}^uk(x,t)u_{x\cdot t}) = V(u_{x\cdot t}) \\ &= ({}^uV(x\cdot t),0)\ \overset{e}{u}_{u_{xt}}\ .\end{aligned}$$

On the other hand,

$$\frac{d}{ds}(V_s \circ K_t({}^uk(x,t)^{-1}u_x))\Big|_{s=0} = \frac{d}{ds}(K_t \circ V_s({}^uk(x,t)^{-1}u_x))\Big|_{s=0}$$

$$= \frac{d}{ds}\,({}^{u}k(x,t)^{-1}K_t(e^{{}^{u}V(x)s}\,u_x))\Big|_{s=0}$$

$$= \frac{d}{ds}\,({}^{u}k(x,t)^{-1}\,e^{{}^{u}V(x)s}\,K_t(u_x))\Big|_{s=0}$$

$$= \frac{d}{ds}\,({}^{u}k(x,t)^{-1}\,e^{{}^{u}V(x)s}\,{}^{u}k(x,t)\,u_{x\cdot t})\Big|_{s=0}$$

$$= ({}^{u}k(x,t)^{-1}\,{}^{u}V(x)\,{}^{u}k(x,t))\,{}^{e}_{u}{}_{u_{xt}}\ .$$

Thus,

$${}^{u}V(x\cdot t) \;=\; {}^{u}k(x,t)^{-1}\,{}^{u}V(x)\,{}^{u}k(x,t). \tag{1.11}$$

Since ${}^{u}V(x\cdot t)$ has the same eigenvalue structure as ${}^{u}V(x)$, $Z^0(x\cdot t) = Z^0(x)$ (t in R). Thus, x is in W_m if and only if xt is in W_m and the result ii follows.

To prove iii we suppose the vector function $Z = (Z^+,Z^-,Z^0)$ is not constant on a homogeneous region Q. Let $c\colon [-r,r] \to Q$ be a path along which Z is not constant and let $t_0 = \inf[t \mid Z(t) \neq Z(-r)]$. Clearly, neither Z^+ nor Z^- can decrease at t_0 since any r^+ or r^- eigenvalue remains so throughout a small neighborhood. Thus, Z^0 decreases at $c(t_0)$. This contradicts the definition of a homogeneous region (Z^0 is constant on Q).

Theorem 1.12. Let H be an r-truncation of a linear lift K of X. On each homogeneous region Q there exist K-invariant vector bundles E^+, E^-, and E^0 such that $TQ = E^+ \oplus E^- \oplus E^0$ and for Y in $E^{\pm}$, $\|K_t(Y)\|$ decays to zero as t tends to $\mp\infty$.

Proof: Let $V_x\colon T_xM \to T_xM$ be the map $Y \to (u^{-1}Y)^{u}V(x)u_x$. It is clear from ${}^{au}V(X) = a\,{}^{u}V(x)a^{-1}$ that V_x is well-defined and that ${}^{u}V(x)$ is the matrix of V_x with respect to the basis u_x. Let $p_1^{s_1}\cdots p_j^{s_j}$ be the minimal polynomial of V_x, where each p_i is a real polynomial of degree one or two. Define E_x^+ to be the kernel of $g^+(V_x)$, where g^+ is the product of those p_i with r^+ eigenvalues.

Since $V_x(bu_x) = b\,{}^uV(x)\,u_x$, we have that $g^+(V_x)(bu_x) = bg^+({}^uV(x))u_x$ and that $g^+(V_x)(bu_x) = 0$ if and only if $bg^+({}^uV(x)) = 0$. This, together with equation 1.11 implies $bu_x \in E^+_x$ if and only if $bg^+({}^uV(x)) = 0$ if and only if $bk(x,t)g^+({}^uV(x))k^{-1}(x,t) = 0$ if and only if $bk(x,t)g^+({}^uV(x)) = 0$ if and only if $bk(x,t)u_{xt} \in E^+_{xt}$. Thus, since $bk(x,t)u_{xt} = K_t(bu_x)$, E^+ is K-invariant.

For bu_x in E^+, there is an r^+-eigenvalue θ (real part $> r$) of V_x and $N > 0$ such that $\|V_t bu_x\| \equiv \|b\exp(t\,{}^uV(x))u_x\| < Ne^{re(\theta)t}$ $(t \le 0)$. Since $V_t = H_{-t}\circ K_t$ and the maximum decay rate of H_{-t} is r, $\|K_t(bu_x)\| \to 0$ as $t \to -\infty$.

It remains to be shown that E^+_x is differentiable in x. This is a local question. Thus, we can consider x to be in R^n rather than M (precompose with a local coordinate map). Let U be any open set with compact closure (which is taken into Q by the local coordinate map). Since $\overline{U}$ is compact, we can find a positively oriented curve C in the right half plane which encloses all r^+ eigenvalues of $[V_x \mid x \text{ in } \overline{U}]$. The E^+_x-projection $P^+(x)$ is given by $-\frac{1}{2\pi i}\int_C (V_x - a)^{-1}\,da$. E^+ is differentiable if and only if P^+ is differentiable ($P^+(R^n) = E^+$ and E^+ has constant dimension) if and only if the matrix of P^+ with respect to u is differentiable (we will denote this matrix by ${}^uP^+$). Since ${}^uV$ is differentiable, the matrix $({}^uV - a)^{-1}$ is differentiable in x for all a on C. Thus, if

$$B(x) = -\frac{1}{2\pi i}\int_C -[{}^uV(x_0) - a]^{-1} D\,{}^uV(x_0)[x-x_0][{}^uV(x_0) - a]^{-1}\,da$$

and

$$A(x) = \frac{\|{}^uP^+(x) - {}^uP^+(x_0) - B(x)\|}{\|x-x_0\|}$$

then

$$A(x) = \frac{\left\| \frac{1}{2\pi i} \int_C [{}^uV(x)-a]^{-1} - [{}^uV(x_0)-a]^{-1} - [{}^uV(x_0)-a]^{-1} D^uV(x_0)[x-x_0][{}^uV(x_0)-a]^{-1} \right\| da}{\|x-x_0\|}$$

$$\leq \frac{1}{2\pi} \int_C \frac{\|[{}^uV(x)-a]^{-1} - [{}^uV(x_0)-a]^{-1} - [{}^uV(x_0)-a]^{-1} D^uV(x_0)[x-x_0][{}^uV(x_0)-a]^{-1}\|}{\|x-x_0\|} da.$$

Since the above integrand tends to 0 as $x \to x_0$, $A(x) \to 0$ as $x \to x_0$, ${}^uP^+$ is differentiable with derivative B.

The arguments for - and 0 are analogous.

§2.

We now consider the special case of the derivative lift of a vector field on M.

Definition 2.1. Given a complete vector field X on M, the derivative lift $\underline{X}$ to TM is given by $\underline{X}_t(Y) = dX_t(\pi Y)[Y]$. The derivative lift of X to LM $\overline{X}$ is given by $(\overline{X}_t(u))^i = \underline{X}_t(u^i)$, $i = 1,\ldots,n$.

Lemma 2.2.

i. The derivative lift of X is a linear lift.

ii. The matrix ${}^uX$ satisfies: $[u^j,X][x]$ = the j-th row of ${}^uX(x)$, written ${}^uX(x)^j$.

Proof: The statement i is trivial. In the proof of ii and in the remainder of the paper, we will use ${}^uk(x,t)$ instead of ${}^ux(x,t)$ to avoid confusion with the base point. Thus, $\overline{X}_t(u_x) = {}^uk(x,t)u_{xt}$. From [1], page 15 we use the formula,

$$\begin{aligned}
[u^j,X][x] &= \lim_{t\to 0} \tfrac{1}{t}(\underline{X}_t(u^j_{x(-t)}) - u^j_x) = \lim_{t\to 0} \tfrac{1}{t}(e_j\overline{X}_t(u_{x(-t)}) - e_j u_x) \\
&= \lim_{t\to 0} \tfrac{1}{t}(e_j{}^uk(x(-t),t)u_x - e_j u_x) = e_j \lim_{t\to 0} \tfrac{1}{t}({}^uk(x(-t),t)-I)u_x \\
&= e_j \tfrac{d}{dt}({}^uk(x(-t),t))\Big|_{t=0} u_x.
\end{aligned}$$

Since $\frac{d}{dt}\,({}^{u}k(x(-t),t))\Big|_{t=0} = \frac{d}{dt}\,({}^{u}k(x,t))\Big|_{t=0} = {}^{u}k(x,0)\,{}^{u}X(x) = {}^{u}X(x)$,

by 1.6 and 1.4, $[u^{j}X](x) = e_j\ {}^{u}X(x)\ u_x = {}^{u}X(x)^{j}u_x$.

Theorem 2.3. <u>Let</u> H <u>be a</u> r-<u>truncation of the derivative lift of</u> X <u>and let</u> u <u>be an orthonormal frame field as defined in</u> 1.5 ii. <u>Let</u> $TQ = E^{+} \oplus E^{-} \oplus E^{0}$ <u>be the vector bundle decomposition on the homogeneous region</u> Q. <u>The map</u>

$$f_u(\underline{b}) = ({}^{u}P^{+}(b)u)_1({}^{u}P^{-}(b)u)_1({}^{u}P^{0}(b)u)_1(x)$$

<u>is a coordinate map from some neighborhood of</u> 0 <u>in</u> R^n <u>onto a neighborhood of</u> x <u>in</u> U.

<u>Proof</u>: The map f_u is the composition of the following differentiable maps:

$$b \to (b,b,{}^{u}P_x^{0}(b)):\quad R^n \to R^{3n}$$

$$(b,b,{}^{u}P_x^{0}(b)) \to (b,b,a(b)):\quad R^{3n} \to R^{2n}\times M,$$

where $a(b) = ({}^{u}P_x^{0}(b)u)_1(x)$

$$(b,b,a(b)) \to (b,{}^{u}P^{-}_{a(b)}(b),a(b)):\quad R^{2n}\times M \to R^{2n}\times M$$

$$(b,{}^{u}P^{-}_{a(b)}(b),a(b)) \to (b,c(b)):\quad R^{2n}\times M \to R^{n}\times M,$$

where $c(b) = ({}^{u}P^{-}_{a(b)}(b)u)_1(a(b))$

$$(b,c(b)) \to ({}^{u}P^{+}_{c(b)}(b),c(b)):\quad R^{n}\times M \to R^{n}\times M$$

$$({}^{u}P^{+}_{c(b)}(b),c(b)) \to ({}^{u}P^{+}_{a(b)}(b)u)_1(c(b)):\quad R^{n}\times M \to M.$$

To show that f_u is regular at x in U, we choose a basis of unit vectors $b_1^{(\ )},\dots,b_{Z^{(\ )}}$ for $u^{-1}(E_x^{(\ )})$, ( ) = +, −, 0. Then, $[b_i^{(\ )}u_x \mid (\ ) = +,-,0$ and $i = 1,\dots,Z_x^{(\ )}]$ is a basis T_xM and

$$df_u(0)b_j^{(\ )} = \frac{d}{dt}(f_u(0+tb_j^{(\ )}))\Big|_{t=0} = \frac{d}{dt}((^u p^{(\ )}(tb_j^{(\ )})u)_1(x))\Big|_{t=0}$$

$$= \frac{d}{dt}((tb_j^{(\ )}u)_1(x))\Big|_{t=0} = \frac{d}{dt}((b_j^{(\ )}u)_t(x))\Big|_{t=0}$$

$$= b_j^{(\ )}u_x .$$

Since u_x is an isomorphism of R^n onto T_xM, f_u is regular at 0. The result follows.

REFERENCES

1. S. Kobayashi and K. Nomizu, Foundations of Differential Geometry, Wiley, 1963.
2. S. Lang, Differential Manifolds, Addison-Wesley, 1972.
3. L. Markus, Continuous matrices and the stability of differential systems, Math. Zeit. 62(1955), pp. 310-319.
4. W. Perrizo, ω-linear vector fields on manifolds, Trans. Amer. Math. Soc. 203(1974), pp. 289-312.
5. S. Smale, Differential dynamical systems, Bull. Amer. Math. Soc. 73(1967).

NORTH DAKOTA STATE UNIVERSITY

INTRODUCTION TO THE CLOSING LEMMA

by

Clark Robinson*

The Closing Lemma states that if a diffeomorphism or flow f has a nonwandering point p_0 then f can be C^1 approximated by a diffeomorphism or flow g which has a periodic orbit through p_0. At first this result was thought to be trivial -- and it is for C^0 approximations. Pugh proved the result for C^1 approximations and it is still unknown for C^2 approximations.

§1. Introduction

The paper of Pugh and Robinson, [4], proves the Closing Lemma for diffeomorphisms, flows, vector fields, Hamiltonian diffeomorphisms and vector fields, and volume preserving diffeomorphisms and vector fields. Unfortunately, the details of the analysis become complicated so it is difficult for the reader to get the main idea of the proof. In this paper, we sketch the proof in the hope that it will give an introduction to the main ideas. We discuss the case of diffeomorphisms. For flows, the proof involves taking transversals to the flow and making perturbations to the Poincaré maps between these transversals much as we do for diffeomorphisms.

§2. Selecting the Orbit

Let p_0 be a nonwandering point for f such that the iterates of p_0, $f^i(p_0)$, are contained in a compact subset of the manifold. Given a $\delta > 0$ there is a point x and an integer $k > 0$ such that $d(p_0,x) < \delta$ and $d(p_0,f^k(x)) < \delta$. The problem is that there may be

*Partially supported by National Science Foundation (MCS77-01080).

some intermediate iterates $f^i(x)$ that are also near to p_0. If we perturb f to g so that $gf^{k-1}(x) = x$ then probably $g^{k-1}(x) \neq f^{k-1}(x)$. This problem adds much of the complexity to the proof.

The first step of the proof applies a selection process to pick an iterate of x through which to construct the periodic orbit. We consider the points $x_i = f^i(x)$ for $0 \leq i \leq k$. If there is a $j = j_1$ with $d(f^i(x),x)$ or $d(f^i(x),f^k(x))$ less than $(2/3)^{1/2}d(x,f^k(x))$ then we consider only $0 \leq i \leq j_1$ or $j_1 \leq i \leq k$ respectively. We repeat this process until it stops. (There are only finitely many points being considered.) By this process we are able to find a point p, an integer n, and a box (an affine map) B such that $d(p_0,p) \leq C\delta$, $d(p_0,f^k(p)) \leq C\delta$ (for some predetermined C), p and $f^n(p)$ are both inside the box B shrunk by a factor of $(3/4)^{1/2}$, and $f^i(p) \notin B$ for $0 < i < n$. For details of the proof which is not hard see [2] or [4, Lemma 4.2].

§3. Closing When There is an Asymptotic Splitting Preserved

The next step is to measure the distance it is possible to move a point x in a small box B with perturbation h of C^1 size ε when h equals the identity outside B. By the mean value theorem, $d(x,h(x))$ is at most ε times the distance of x to the boundary of the box. If B is much shorter in one direction than the others then h can move x a distance proportional to the shortest side of B, L_1, but not proportional to the longer sides:

$$d(x,h(x)) \leq \varepsilon d(x,\partial B) \approx \varepsilon\, 1/2[1-(3/4)^{1/2}]L_1.$$

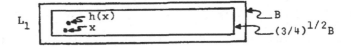

Thus to move a point a distance proportional to L_1, we need about $1/\varepsilon$ pushes. To move a distance of L_2 we would need about $L_2/\varepsilon L_1$ pushes which becomes large as L_2/L_1 becomes large.

As we apply repeated perturbations along the orbit of p, the shape of the boxes $f^i(B)$ may change. For simplicity let us assume that the manifold is two dimensional, that f is linear (affine) along the orbit of p, and that f preserves the vertical and horizontal directions. Moreover, let us assume that the horizontal direction asymptotically grows more rapidly than the vertical direction. In the fundamental lemma which selects p and $q = f^n(p)$, it turns out to be possible to choose the box B so that it is very tall and slim. In fact we can take B so that for $0 \le i \le N = 40/\varepsilon$, $f^i(B)$ is taller than it is wide where $N = 40/\varepsilon$ is the number of pushes we need in each direction. Since the horizontal direction eventually stretches more than the vertical direction we can find n_2 such that $f^i(B)$ is wider than it is tall for $n_2 \le i \le n_2 + N$.

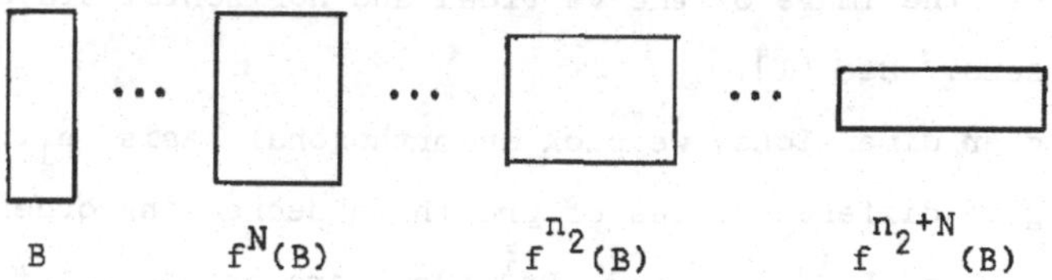

We then construct a perturbation h which equals the identity outside

$$\cup \{f^i(B) : 1 \le i \le N \text{ or } n_2 \le i \le n_2 + N\}.$$

The first N perturbations allow us to get $g = h \circ f$ such that $g^N(q)$ has moved so it has the same horizontal coordinate as $f^N(p)$. The

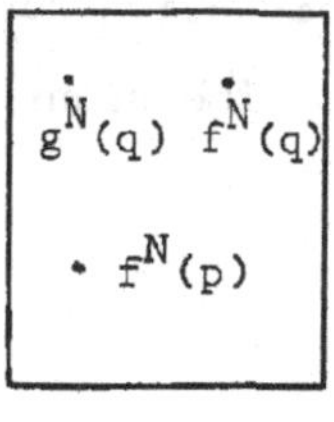

next N perturbations allow us to move vertically so that

$$g^{n_2+N}(q) = f^{n_2+N}(p).$$

For $n_2 + N \leq i \leq n$, $f^i(p)$ is outside the support of $h = g \circ f^{-1}$ so

$$g^n(q) = g^{n-n_2-N} f^{n_2+N}(p) = f^n(p) = q.$$

Thus q is a periodic orbit for g. Another perturbation allows us to move the periodic orbit over through p_0.

Above we talked as if f^i were linear. The nonlinearities are taken care of in the proof by taking B very small so that for $0 \leq i \leq n_2 + N$, $f^i(B)$ is very near $f^i(p_0)$ and so $f^i(x)$ is C^1 near to $f^i(p_0) + Df^i(p_0)(x-p_0)$. In [4] this is made precise by introducing a certain type of double limit to show it is enough to consider the linear case.

This much of the proof is very nearly what Pliss needed for his work on the converse to the structural stability theorem. There the angle between the image of the vertical and horizontal stayed bounded away from zero. See [1].

In higher dimensions, we pick an orthogonal basis $e_1, \ldots, e_m \in T_pM$ that give different rates of growth in decreasing order. Here we also assume that $Df^i(p)e_j$ and $Df^i(p)e_k$ are orthogonal for $0 \leq i \leq n$. We can get p and $q = f^n(p)$ near p_0, a box B, and $0 = n_1 < N < n_2 < n_2 + N < \ldots < n_{m-1} + N < n_m < n_m + N < n$ such that (i) p and q are inside the box $(3/4)^{1/2}B$, (ii) $f^i(p) \notin B$ for $0 < i < n$, and (iii) for $n_j \leq i \leq n_j + N$, $Df^i(p)e_j$ is the shortest side of the box $f^i(B)$. We construct the perturbation in $f^i(B)$ so that we push in the jth direction for $n_j \leq i \leq n_j + N$. The rest of the proof is similar to two dimensions.

§4. Shear

Finally, we discuss briefly the effect due to the sides of the box not remaining orthogonal. We can find a basis of vectors $e_1,\ldots,e_m \in T_pM$ with decreasing asymptotic rates of growth. Assume $Df^i(p)e_m$ is the shortest side. The distance from the points of interest, x, to the boundary, $\partial f^i(B)$, is no longer proportional to the

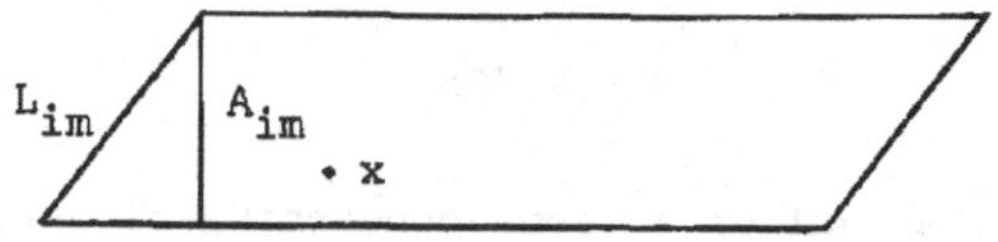

length of the side L_{im} but to the length of the altitude A_{im}. It turns out to be necessary to push in the direction of the altitude, perpendicular to span $\{Df^i(p)e_j : 1 \le j \le m-1\}$. If the sides in the jth directions for $1 \le j \le m-1$ are much longer than the mth altitude, the contribution of the push is mainly in the mth direction. The contribution in the first direction, for example, can easily be overcome when we push in that direction when it is the shortest side. For details about the linear algebra of this part of the proof see [3] or [4].

§5. Remarks on the Order of Choices

The first step in the proof is to analyze the linear maps $Df^i(p_0)$, the various rates of growth, and the altitude maps. If two directions e_j and e_{j+1} grow at the same asymptotic rate then there is a bound, b, called the bolicity, on the lack of conformality in the plane for all the linear maps $Df^i(p_0)$. In terms of this bound, we know that we need at most $N = 40b/\varepsilon$ pushes in each direction to close the orbit.

Next, we select the n_j and ratios of the length of the sides of the box B, $\lambda_1,\ldots,\lambda_m$, so that for $n_j \le i \le n_j + N$ the jth altitude of the box given by $\{Df^i(p_0)\lambda_k e_k : 1 \le k \le m\}$ is smaller than the other lengths of the sides. Next, we choose a very small neighborhood U of p_0 so the f^i is very nearly linear on U for $0 \le i \le n_m + N$. The fundamental lemma then selects p and $q = f^n(p)$ inside U with $n \ge n_m + N$ and the box B with sides of lengths proportional to $\lambda_1,\ldots,\lambda_m$, $p,q \in (3/4)^{1/2}B$, and $f^i(p) \notin B$ for $0 < i < n$. We then proceed as discussed above.

REFERENCES

1. V.A. Pliss, A variant of a lemma concerning closure, Diff. Equat. 7(1971), 642-650.

2. C. Pugh, The closing lemma, Amer. J. Math. 89(1967), 956-1021.

3. C. Pugh, On arbitrary sequences of isomorphisms of R^m, Trans. Amer. Math. Soc. 184(1973), 387-400.

4. C. Pugh and C. Robinson, The C^1 closing lemma including Hamiltonians, preprint.

NORTHWESTERN UNIVERSITY

ON THE PSEUDO ORBIT TRACING PROPERTY
AND ITS RELATIONSHIP TO STABILITY

by

Peter Walters

§0. Introduction

Rufus Bowen has stated that the tracing of pseudo orbits is the most important dynamical property of Axiom A maps. ([1]). At the end of his paper [3] he sketches how this property can be used to prove stability results. In Section 2 of this article we give the details of how the pseudo orbit tracing property for an expansive homeomorphism implies certain stability results. These results give the known theorems for Anosov diffeomorphisms when they are combined with Bowen's result that Anosov diffeomorphisms have the pseudo orbit tracing property. We also show that subshifts of finite type have a stability property that is analogous to structural stability. Bill Parry is responsible for this idea. In Section 4 we show that topologically stable homeomorphisms on manifolds of dimension ≥ 2 have the pseudo orbit tracing property. Dennis Sullivan informed us of Lemma 10. This work was largely carried out in the autumn of 1975 at I.H.E.S., France, where the author benefited from conversations with Rufus Bowen and Dennis Sullivan. We mentioned these results to A. Moromoto when he visited the University of Warwick in the summer of 1976 and he has incorporated them into two of his papers [5], [6].

§1. The Pseudo Orbit Tracing Property

X will always denote a compact metric space with metric d and $T: X \to X$ will always be a homeomorphism.

If $\varphi: X \to X$, $\psi: X \to X$ are continuous then $d(\varphi,\psi)$ will denote $\sup_{x \in X} d(\varphi(x),\psi(x))$.

The dynamical study of T is largely concerned with the orbits, $\{T^n(x)\}_{n=-\infty}^{+\infty}$, of T. If $S: X \to X$ is a homeomorphism with $d(S,T) < \delta$ (i.e. S is a perturbation of T) then each orbit, $\{S^n(x)\}_{-\infty}^{\infty}$, of S is almost an orbit of T in the same sense that $d(T(S^n(x)), S^{n+1}(x)) < \delta \ \forall \ n \in Z$. This can be used to motivate the definition of a pseudo-orbit of T.

<u>Definition 1</u>. ([2], [3]). Let $\delta > 0$. A δ pseudo-orbit for $T: X \to X$ is a bisequence $\{x_n\}_{-\infty}^{\infty}$ of points of X such that $d(T(x_n), x_{n+1}) < \delta \ \forall \ n \in Z$.

For T to be "stable" we would like each pseudo-orbit for T to be closely related to an actual orbit of T.

<u>Definition 2</u>. ([2], [3]). A δ pseudo-orbit, $\{x_n\}_{-\infty}^{\infty}$, for T is ε-traced by x if $d(x_n, T^n(x)) < \varepsilon \ \forall \ n \in Z$.

<u>Definition 3</u>. ([2], [3]). T is said to have the pseudo-orbit tracing property (P.O.T.P.) if $\forall \ \varepsilon > 0 \ \exists \ \delta > 0$ such that each δ pseudo-orbit for T is ε-traced by some point of X.

It is clear that the statement "T has the P.O.T.P." does not depend on the choice of metric on X, and it is preserved under topological conjugacy.

To get a feeling for this definition let us consider which shift systems have the P.O.T.P. Let k be a fixed natural number and let $C = \{0,1,\ldots,k-1\}$. Put the discrete topology on C. Consider the product space $\Sigma = \prod_{-\infty}^{\infty} C$, equipped with the product topology, and the shift homeomorphism $\sigma: \Sigma \to \Sigma$ defined by $(\sigma(w))_n = w_{n+1}$, where $w = (w_n)_{-\infty}^{\infty}$. A metric on X is defined by $d(x,y) = 2^{-m}$ if m is the largest natural number with $x_n = y_n \ \forall \ |n| < m$, and $d(x,y) = 1$ if $x_0 \neq y_0$. If X is a closed subset of Σ with $\sigma X = X$ then $\sigma|_X : X \to X$ is called a subshift. We usually write this as $\sigma: X \to X$. A subshift $\sigma: X \to X$ is said to be of finite type if there exists some natural number N and a collection of blocks of length $N+1$ with the property that $x = (x_n)_{-\infty}^{\infty} \in X$ if and only if each block $(x_i, \ldots, x_{i+N})$ in x of length $N+1$ is one of the prescribed blocks. The least such natural number N is called the order of the subshift of finite type. Every subshift of finite type is topologically conjugate to one of order 1 (but the number of symbols k changes). One just takes a new symbol space C' consisting of the allowable blocks of length N, and then the conjugacy is given by

$$\varphi(\ldots x_{-1}\overset{*}{x}_0 x_1 \ldots) = \left(\ldots \begin{pmatrix} x_{-1} \\ \vdots \\ x_{N-2} \end{pmatrix} \begin{pmatrix} \overset{*}{x}_0 \\ \vdots \\ x_{N-1} \end{pmatrix} \begin{pmatrix} x_1 \\ \vdots \\ x_N \end{pmatrix} \ldots \right).$$

Theorem 1. Let $\sigma\colon X \to X$ be a subshift. $\sigma\colon X \to X$ has the P.O.T.P. if and only if it is of finite type.

Proof. Suppose $\sigma\colon X \to X$ is of finite type. By the above remarks we can assume it is of order 1. Let $\varepsilon > 0$ be given. Choose $m \geq 1$ with $2^{-m} < \varepsilon$. Then if $x = (x_n)$, $y = (y_n)$ and $d(x,y) < 2^{-(m+1)}$ we have $x_i = y_i$ when $|i| < m$. In particular $x_0 = y_0$. Let $\{x^{(j)}\}_{j=-\infty}^{\infty}$ be a $2^{-(m+1)}$ pseudo-orbit for σ. Since $d(\sigma x^{(j)}, x^{(j+1)}) < 2^{-(m+1)}$ we have $x_1^{(j)} = x_0^{(j+1)}$ $\forall\, j$. Consider $x = (x_n)$ where $x_n = x_0^{(n)}$. Then all blocks in x of length 2 are allowable because $(x_0^{(n)}, x_0^{(n+1)}) = (x_0^{(n)}, x_1^{(n)})$. Therefore $x \in X$. Also $(\sigma^n x)_j = x_j^{(n)}$ for $|j| < m$ and so $d(\sigma^n x, x^{(n)}) \leq 2^{-m} < \varepsilon$. We have shown σ has the P.O.T.P.

Conversely suppose $\sigma\colon X \to X$ has the P.O.T.P. Choose δ to correspond to $\varepsilon = 1/2$ in Definition 3. Choose N so that $2^{-N} < \delta$. We shall show σ is a subshift of finite type of order at most $2N+1$. It will suffice to show that for each $m \geq 2N+1$, every block $(a_1, \ldots, a_m)$ in Σ which has all of its subblocks of length $2N+1$ allowable in X is itself allowable in X. We shall use induction on

m. The statement is clearly true for $m = 2N+1$. Suppose now that each block of length m which has all of its subblocks of length $2N+1$ allowable in X is itself allowable in X. Suppose $(a_1,\ldots,a_{m+1})$ is a block of length $m+1$ and suppose all of its subblocks of length $2N+1$ are allowable in X. By the inductive assumption there exists $x = (x_n) \in X$ with $x_{-N-1+i} = a_i$, $1 \le i \le m$, and $y = (y_n) \in X$ with $y_{-N-1+i} = a_i$, $2 \le i \le m+1$. We have $d(x,y) \le 2^{-N} < \delta$ so that $\{\ldots,\sigma^{-2}x,\sigma^{-1}x,x,y,\sigma y,\sigma^2 y,\ldots\}$ is a δ pseudo-orbit for σ. Let $z \in X$ be a point which 1/2-traces this pseudo-orbit. Then $z_{-N-1+i} = a_i$, $1 \le i \le m+1$, and so $(a_1\ldots a_{m+1})$ is allowable in X.

Hence the P.O.T.P. picks out those subshifts with the most desirable dynamical properties: every 0-dimensional basic set of an Axiom A diffeomorphism is topologically conjugate to a subshift of finite type; and subshifts of finite type are exactly the subshifts that have canonical coordinates. ([2], [4]).

Most of the time we shall be dealing with a homeomorphism $T: X \to X$ which is expansive.

<u>Definition 4</u>. A homeomorphism $T: X \to X$ is expansive if $\exists\, e(T) > 0$ such that if $d(T^n(x), T^n(y)) \le e(T)$ $\forall\, n \in Z$ then $x = y$. Such numbers $e(T)$ are called expansive constants.

The possible values the expansive constant can take depends on the metric d but the notion of expansiveness is independent of the metric.

Any subshift is expansive: if we take the metric defined by $d(x,y) = 2^{-m}$ if m is the largest with $x_n = y_n$ $\forall\, |n| < m$, then 1/2 is an expansive constant because $d(\sigma^n(x), \sigma^n(y)) \le 1/2$ implies $(\sigma^n(x))_0 = (\sigma^n(y))_0$, i.e. $x_n = y_n$.

An expansive homeomorphism determines the topology of X in the following sense.

Lemma 2. *Let* $T: X \to X$ *be an expansive homeomorphism with expansive constant* $e(T)$ *relative to the metric* d. *For each* $N \geq 1$ *there exists* $\delta > 0$ *with the property that* $d(x,y) < \delta$ *implies* $d(T^n(x), T^n(y)) \leq e(T)$ *for all* n *with* $|n| < N$. *Conversely, given* $\varepsilon > 0$ *there exists* $N \geq 1$ *such that* $d(T^n(x), T^n(y)) \leq e(T)$ *for all* n *with* $|n| < N$ *implies* $d(x,y) < \varepsilon$.

Proof. Let N be given. The continuity of T shows that $\delta > 0$ can be chosen as in the statement.

Conversely, let $\varepsilon > 0$ be given. If no N can be chosen with the property stated then for each $N \geq 1$ there exists x_n, y_n X with $d(T^n(x_n), T^n(y_n)) \leq e(T)$ for all n with $|n| < N$ and $d(x_n, y_n) \geq \varepsilon$. Choose N_i with $x_{n_i} \to x$ and $y_{n_i} \to y$. Then $d(x,y) \geq \varepsilon$ and also $d(T^n(x), T^n(y)) \leq e(T)$ for all n. This contradicts the expansive property of T.

Lemma 3. *Let* $T: X \to X$ *be an expansive homeomorphism with the P.O.T.P. If* $\varepsilon < e(T)/2$ *and* δ *corresponds to* ε *as in Definition 3 then there is a unique* $x \in X$ *which* ε*-traces a given* δ *pseudo-orbit.*

Proof. If y also ε-traces $\{x_n\}$ then

$$d(T^n(x), T^n(y)) \leq d(T^n(x), x_n) + d(x_n, T^n(y)) < 2\varepsilon < e(T) \quad \forall\, n \in Z.$$

§2. Topological Stability

The following definition is taken from [11].

Definition 5. A homeomorphism $T: X \to X$ is topologically stable if $\forall\, \varepsilon > 0\ \exists\, \delta > 0$ so that if $S: X \to X$ is any homeomorphism with $d(S,T) < \delta$ then there is a continuous map $h: X \to X$ with $hS = Th$ and

$d(h,id) < \varepsilon$.

The ideas for the proof of the following theorem are in [3].

Theorem 4. <u>An expansive homeomorphism</u> $T: X \to X$ <u>with the P.O.T.P. is topologically stable. (In fact we shall show the stronger stability statement: Let</u> $e(T)$ <u>be an expansive constant for</u> T. <u>Then</u> $\forall\ \varepsilon > 0$ <u>with</u> $\varepsilon < e(T)/3$, $\exists\ \delta > 0$ <u>so that if</u> $S: X \to X$ <u>is a homeomorphism and</u> $d(S,T) < \delta$ <u>then there is a unique continuous</u> $h: X \to X$ <u>with</u> $hS = Th$ <u>and</u> $d(h,id) < \varepsilon$.)

<u>Proof</u>. Let $\varepsilon < e(T)/3$ and choose δ to correspond to ε as in Definition 3. Let $S: X \to X$ be a homeomorphism with $d(S,T) < \delta$. Let $x \in X$. The S-orbit, $\{S^n(x)\}_{-\infty}^{\infty}$, is a δ-pseudo-orbit for T because $d(TS^n(x), S^{n+1}(x)) \le d(T,S) < \delta$. By Lemma 3 there is a unique point $h(x) \in X$ whose T-orbit traces $\{S^n(x)\}_{-\infty}^{\infty}$. This defines a map $h: X \to X$ with $d(T^n h(x), S^n(x)) < \varepsilon \quad \forall\ n \in Z, \quad \forall\ x \in X$. Putting $n = 0$ gives $d(h,id) < \varepsilon$.

Since $d(T^n hS(x), S^{n+1}(x)) < \varepsilon\ \ \forall\ n \in Z$ and $d(T^n(Th(x)), S^{n+1}(x)) = d(T^{n+1}h(x), S^{n+1}(x)) < \varepsilon\ \ \forall\ n \in Z$ we have that both $hS(x)$ and $Th(x)$ ε-trace $\{S^{n+1}(x)\}_{-\infty}^{\infty}$. By Lemma 3 we have $hS = Th$.

We now show that h is continuous. Let $\lambda > 0$ be given. Using Lemma 2 choose N so that $d(T^n(u), T^n(v)) < e(T)\ \ \forall\ |n| \le N$ implies $d(u,v) < \lambda$. Choose $\eta > 0$ such that $d(x,y) < \eta$ implies $d(S^n(x), S^n(y)) < e(T)/3\ \ \ \forall\ |n| \le N$. Then if $d(x,y) < \eta$

$$\begin{aligned} d(T^n h(x), T^n h(y)) &= d(hS^n(x), hS^n(y)) \\ &\le d(hS^n(x), S^n(x)) + d(S^n(x), S^n(y)) + d(S^n(y), hS^n(y)) \\ &< \varepsilon + \frac{e(T)}{3} + \varepsilon \qquad \forall\ |n| \le N \\ &< e(T) \qquad \forall\ |n| \le N. \end{aligned}$$

Therefore $d(x,y) < \eta$ implies $d(h(x), h(y)) < \lambda$, and the continuity of h is proved. The map h is the only one with $hS = Th$ and $d(h,id) < \varepsilon$ since if ℓ is another one,

$$\begin{aligned} d(T^n\ell(x), T^nh(x)) &= d(\ell S^n(x), hS^n(x)) \\ &\le d(\ell S^n(x), S^n(x)) + d(S^n(x), hS^n(x)) \\ &< 2\varepsilon < e(T) \quad \forall n \in Z, \end{aligned}$$

and so $\ell(x) = h(x)$.

Remark. If X is a compact manifold and ε is sufficiently small then $d(h,id) < \varepsilon$ implies that h maps X onto X ([7], p. 36). In general spaces this need not be so. Consider the shift $\sigma: \Sigma \to \Sigma$ where $\Sigma = \prod_{-\infty}^{\infty} C$ and $C = \{0,1\}$. Define $S: \Sigma \to \Sigma$ by $(S(x))_n = x_n$ if $n < -m$ or $n > m$, $(S(x))_n = x_{n+1}$ if $-m \le n < m$, $(S(x))_m = x_{-m}$. Then $d(S,\sigma) = 1/2^m$, using the metric d on Σ introduced in Section 1, so if m is sufficiently large we have $hS = \sigma h$ for some continuous $h: X \to X$ with $d(h,id) < \varepsilon$. Note that $S^{2m+1} = id$ so that $\sigma^{2m+1}h(x) = h(x)$ $\forall x \in \Sigma$. Therefore the image of h is contained in the finite subset of Σ consisting of the periodic points of order $2m+1$. We shall consider later conditions on the perturbation S of the shift σ to ensure that the conjugating map h is a homeomorphism. A general result in this direction is:

Theorem 5. Let $T: X \to X$ be an expansive homeomorphism with the P.O.T.P. If the perturbation $S: X \to X$ of T (in the statement of Theorem 4) is also assumed to be expansive with an expansive constant $e(S) \ge 2\varepsilon$ then the corresponding conjugating map h is injective.

Proof. Let $h(x) = h(y)$. Then

$$\begin{aligned} d(S^n(x),S^n(y)) &\le d(S^n(x),hS^n(x)) + d(hS^n(x),hS^n(y)) + d(hS^n(y),S^n(y)) \\ &= d(S^n(x),hS^n(x)) + d(T^nh(x),T^nh(y)) + d(hS^n(y),S^n(h)) \end{aligned}$$

$$< \; 2\varepsilon \; \leq e(S).$$

Hence $x = y$.

§3. Applications of Section 2

1. Anosov diffeomorphisms

Suppose $T: M \to M$ is an Anosov diffeomorphism of a compact manifold. Bowen has shown that T has a P.O.T.P. ([2], p. 74). Any Anosov diffeomorphism is also expansive. ([8], p. 108). These results and Theorem 4 give the following result of the author: An Anosov diffeomorphism is topologically stable. ([11]).

We shall now indicate how the celebrated theorem of Anosov -- that an Anosov diffeomorphism is structurally stable -- can be deduced from Theorem 5. Let $S: M \to M$ be a diffeomorphism which is C' close to $T: M \to M$. Let $B(M,M)$ denote the Banach manifold of all maps between M. Then the map of $B(M,M)$ defined by $\varphi \to S\circ\varphi\circ T^{-1}$ is C' close to the map $\varphi \to T\circ\varphi\circ T^{-1}$ in a neighborhood of $id \in B(M,M)$. Since id is a hyperbolic fixed point of the second map, these maps are locally topologically conjugate near id. It follows from this that S is expansive and if S is sufficiently C' close to T then $e(S) > e(T)/2$. Then Theorem 5 and the remark that preceeds it imply the structural stability of T.

2. Subshifts of finite type

It follows from Theorems 1 and 4 that any subshift of finite type is topologically stable. We now show that subshifts of finite type have a structural stability property if we require the perturbations to be close in a sense analogous to a C' metric. This work was done with Bill Parry who suggested this type of theorem.

Theorem 6. *Let* $T: X \to X$ *be a topological transitive subshift of finite type of order* 1 *and let* d *be the metric on* X *defined by*

$d((x_n),(y_n)) = 1/2^N$ <u>if</u> N <u>is the largest integer with</u> $x_n = y_n$ <u>for all</u> $|n| < N$. <u>There exists</u> $\varepsilon_0 > 0$ <u>so that</u> $\forall\ \varepsilon < \varepsilon_0$, $\varepsilon > 0$, $\exists$ $\delta > 0$ <u>such that if</u> $S: X \to X$ <u>is any homeomorphism with</u> $d(S,T) < \delta$, $|d(Sx,Sy) - d(Tx,Ty)| \le \delta d(x,y)$ <u>and</u> $|d(S^{-1}x,S^{-1}y) - d(T^{-1}x,T^{-1}y)| \le \delta d(x,y)$ $\forall\ x,y \in X$, <u>then there exists a unique homeomorphism</u> $h: X \to X$ <u>with</u> $hS = Th$ <u>and</u> $d(h,id) < \varepsilon$.

<u>Proof</u>. We know that T is expansive and satisfies the P.O.T.P. so we know T is topologically stable. It suffices to show the conjugating map $h: X \to X$ is injective and surjective.

We shall show h is injective by using Theorem 5. Our metric d has the property that if $x \ne y$ and $d(x,y) \le 1/2$ then either $d(Tx,Ty) = 2d(x,y)$ or $d(T^{-1}x,T^{-1}y) = 2d(x,y)$. If S is as in the statement of the theorem then if $x \ne y$ and $d(x,y) \le 1/2$ we have either $d(Sx,Sy) \ge (2-\delta)d(x,y)$ or $d(S^{-1}x,S^{-1}y) \ge (2-\delta)d(x,y)$. Suppose $x \ne y$ are so that $d(S^nx,S^ny) < 1/2$ $\forall\ n \in Z$. We can suppose without loss of generality that $d(Sx,Sy) \ge (2-\delta)d(x,y)$ (rather than $d(S^{-1}x,S^{-1}y) \ge (2-\delta)d(x,y)$.) Then we must have $d(S^2x,S^2y) \ge (2-\delta)d(Sx,Sy)$ because the other possiblility is that $d(x,y) \ge (2-\delta)^2 d(x,y)$ which is impossible. We claim that $d(S^nx,S^ny) \ge (2-\delta)d(S^{n-1}x,S^{n-1}y)$ $\forall\ n \ge 1$. Suppose this is true for $n \le k$. Then either it holds for $n = k+1$ or else $d(S^{k-1}x,S^{k-1}y) \ge (2-\delta)^2 d(S^{k-1}x,S^{k-1}y)$ which is impossible. Therefore we have $1/2 \ge (2-\delta)^n d(x,y)$ $\forall\ n \ge 1$ and this is impossible. We conclude that $x = y$ and that S is expansive with $1/2$ as an expansive constant. Theorem 5 gives us that h is injective.

Let $\alpha = \{A_0,\dots,A_{k-1}\}$ denote the natural partition of X, i.e. $A_i = \{x \in X \mid x_0 = i\}$. Each A_i is an open and closed subset of X. We shall show $\vee_{-n}^{n} S^i\alpha = \vee_{-n}^{n} T^i\alpha$ for each $n \ge 0$. Let $\varphi = ST^{-1}$. We have

$$\delta d(x,y) \ \ge\ |d(Sx,Sy) - d(Tx,Ty)| \ =\ |d(\varphi Tx,\varphi Ty) - d(Tx,Ty)|$$

so that $|d(\phi u,\phi v)-d(u,v)| \leq \delta d(T^{-1}u,T^{-1}v) \leq 2\delta d(u,v)$. But this gives $\left|\frac{d(\phi u,\phi v)}{d(u,v)} - 1\right| \leq 2\delta$ if $u \neq v$ and since $\frac{d(\phi u,\phi v)}{d(u,v)}$ takes values in the set $\{2^n\}_{-\infty}^{\infty}$ we have that ϕ is an isometry. Therefore $\phi(V^n_{-n}\, T^i\alpha) = V^n_{-n}\, T^i\alpha$. This implies $S^{-1}(V^n_{-n}\, T^i\alpha) = V^{n-1}_{-(n+1)}\, T^i\alpha$. Replacing S by S^{-1} and T by T^{-1} in the above reasoning we also get $S(V^n_{-n}\, T^i\alpha) = V^{n+1}_{-(n-1)}\, T^i\alpha$.

We can now show $V^n_{-n}\, S^i\alpha = V^n_{-n}\, T^i\alpha$ by inductions on n. It is clearly true for $n = 0$. Assume it is true for $n = N$. Then $V^{N-1}_{-(N+1)}\, S^i\alpha = S^{-1}(V^N_{-N}\, S^i\alpha) = S^{-1}(V^N_{-N}\, T^i\alpha) = V^{N-1}_{-(N+1)}\, T^i\alpha$ and $V^{N+1}_{-(N+1)}\, S^i\alpha = S(V^N_{-N}\, S^i\alpha) = S(V^N_{-N}\, T^i\alpha) = V^{N+1}_{-(N-1)}\, T^i\alpha$. These imply $V^{N+1}_{-(N+1)}\, S^i\alpha = V^{N+1}_{-(N+1)}\, T^i\alpha$.

Since α is a topological generator for T we have that the topological entropy of T is given by $h(T) = h(T,\alpha)$. ([13], p. 170). Therefore $h(S) \geq h(S,\alpha) = h(T,\alpha) = h(T)$. Since $S: X \to X$ is expansive it has a measure μ of maximal entropy (i.e., $h_\mu(S) = h(S)$) ([2], p. 65). The measure $\mu' = \mu o h^{-1}$ is supported on a subset of $h(X)$ and $h_{\mu'}(T) = h_\mu(S) \geq h(T)$. Therefore μ' is a measure of maximal entropy for T. However since T is topologically transitive it has a unique measure m with maximal entropy and m is supported on the whole of X. ([10]). Therefore $m = \mu'$ and $h(X) = X$.

§4. When Does Topological Stability Imply the P.O.T.P.?

We now consider the problem of when a topologically stable homeomorphism has the P.O.T.P. To do this we shall need to be able to choose a homeomorphism, close to id, to map a given finite set of points to a nearby given set of points (Lemma 10). Dennis Sullivan showed us how to choose such a homeomorphism on a compact smooth manifold. The reason we only have to specify finite sets is the following.

Lemma 8 ([2]), p. 75). *Suppose the homeomorphism* $T: X \to X$ *has the following tracing property for finite pseudo-orbits:* $\forall\ \varepsilon > 0 \exists\ \delta > 0$ *such that if the points* $\{x_0, \ldots, x_k\}$ *satisfy* $d(Tx_n, x_{n+1}) < \delta$, $0 \leq n \leq k-1$ *then there exists* $x \in X$ *with* $d(T^n x, x_n) < \varepsilon$, $0 \leq n \leq k-1$. *Then* T *has the P.O.T.P.*

Proof. Let $\varepsilon > 0$ be given. Choose δ as in the statement of the lemma.

Let $\{x_n\}^{\infty}_{\infty}$ be a δ pseudo-orbit for T. For each $m > 0$ there is $x^{(m)} \in X$ with $d(T^n z^{(m)}, x_{n-m}) < \varepsilon$, $0 \leq n \leq 2m$. Let $w^{(m)} = T^m z^{(m)}$. Then $d(T^j w^{(m)}, x_j) < \varepsilon$, $|j| \leq m-1$. Choose a convergent subsequence $w^{(m_i)} \to w$. Then $d(T^j w, x_j) \leq \varepsilon\ \forall\ j \in Z$ and therefore w 2ε-traces $\{x_n\}_{-\infty}^{\infty}$.

Lemma 9. Let $T: M \to M$ be a homeomorphism of a compact manifold. Let $k \geq 0$ be an integer and let $\tau > 0$ and $\eta > 0$ be given. Then for any set of points $\{x_0, x_1, \ldots, x_k\}$ with $d(T(x_i), x_{i+1}) < \tau$ $(0 \leq i \leq k-1)$ there exists a set of points $\{x_0', x_1', \ldots, x_k'\}$ such that

(a) $d(x_i, x_i') < \eta$ $\quad (0 \leq i \leq k)$

(b) $d(T(x_i'), x_{i+1}') < 2\tau$ $\quad (0 \leq i \leq k-1)$

(c) $x_i' \neq x_j'$ if $i \neq j$ $\quad (0 \leq i \leq k, \ 0 \leq j \leq k)$.

Proof. We use induction on k. For $k = 0$ the statement is true. Suppose the lemma is true for $k-1$ and we shall prove it for k. Let $\tau > 0$ and $\eta > 0$ be given. We can suppose $\eta < \tau$. Choose $\lambda > 0$ such that $d(x,y) < \lambda$ implies $d(Tx,Ty) < \tau$ and $\lambda > \eta$. Let $\{x_0, \ldots, x_k\}$ be given so that $d(T(x_i), x_{i+1}) < \tau$, $0 \leq i \leq k-1$. By assumption we can choose $\{x_0', \ldots, x_{k-1}'\}$ so that $d(x_i, x_i') < \lambda$ $(0 \leq i \leq k-1)$, $d(T(x_i'), x_{i+1}') < 2\tau$ $(0 \leq i \leq k-2)$, and $x_i' \neq x_j'$ if $i \neq j$ $(i \leq k-1, \ j \leq k-1)$. We know $d(T(x_{k-1}'), x_k) \leq d(T(x_{k-1}'), T(x_{k-1})) + d(T(x_{k-1}), x_k) < 2\tau$, so choose x_k' so that $x_k' \neq x_j'$ if $j \leq k-1$, $d(x_k', x_k) < \eta$, and $d(T(x_{k-1}'), x_k') < 2\tau$.

Lemma 10 ([9], Lemma 13). Let M be a compact manifold of dimension ≥ 2. Suppose a finite collection $\{(p_i, q_i) \in M \times M \mid i = 1, \ldots, r\}$ is specified together with a small $\lambda > 0$ such that

(i) $d(p_i, q_i) < \lambda$ all i, and

(ii) if $i \neq j$ then $p_i \neq p_j$ and $q_i \neq q_j$.

Then there exists a diffeomorphism $f: M \to M$ such that

(a) $d(f, id) < 2\pi\lambda$, and

(b) $f(p_i) = q_i$ $\quad (1 \leq i \leq r)$.

Theorem 11. Let $T: M \to M$ be a topologically stable homeomorphism of a compact manifold of dimension ≥ 2. Then T has the P.O.T.P.

Proof. We shall verify the condition of Lemma 8.

Let $\varepsilon > 0$ be given and let δ correspond to ε as in the definition of topological stability. Suppose $\{x_0, x_1, \ldots, x_k\}$ be so that $d(T(x_i), x_{i+1}) < \delta/4\pi$ for $0 \leq i \leq k-1$. By Lemma 9 there exists $\{x'_0, x'_1, \ldots, x'_k\}$ such that $d(x_i, x'_i) < \varepsilon$ $(0 \leq i \leq k)$, $d(T(x'_i), x'_{i+1}) < \delta/2\pi$ $(0 \leq i \leq k-1)$, $x'_i \neq x'_j$ if $i \neq j$ $(i < k, j < k)$, and $T(x'_i) \neq T(x'_j)$ if $i \neq j$ $(i \leq k-1, j \leq k-1)$. By Lemma 10 there is a homeomorphism $f: M \to M$ with $d(f, id) < \delta$ and $fT(x'_i) = x'_{i+1}$ $(0 \leq i \leq k-1)$. Let $S = f \circ T$. Then $d(S,T) < \delta$ and $S(x'_i) = x'_{i+1}$ $(0 \leq i \leq k-1)$. By topological stability there is a continuous map $h: M \to M$ such that $d(h, id) < \varepsilon$ and $hS = Th$. Then

$$\begin{aligned} d(T^i h(x'_0), x_i) &= d(hS^i(x'_0), x_i) = d(h(x'_i), x_i) \\ &\leq d(h(x'_i), x'_i) + d(x'_i, x_i) \\ &< \varepsilon + \varepsilon = 2\delta \qquad (0 \leq i \leq k-1). \end{aligned}$$

Hence for each set $\{x_0, x_1, \ldots, x_k\}$ with $d(T(x_i), x_{i+1}) < \delta/4\pi$ $(0 \leq i \leq k-1)$ there exists $y \in X$ such that $d(T^i(y), x_i) < 2\varepsilon$ $(0 \leq i \leq k-1)$. It follows from Lemma 8 that T has the P.O.T.P.

REFERENCES

1. R. Bowen, On Axiom A diffeomorphisms, this volume.

2. R. Bowen, Equilibrium States and the Ergodic Theory of Axiom A Diffeomorphisms, Springer Lecture Notes, Vol. 470, 1975.

3. R. Bowen, w-limit sets for Axiom A diffeomorphisms, J. Diff. Eqns. 18(1975), 333-339.

4. R. Bowen, Topological entropy and Axiom A, Proc. Symp. Pure Math. 14(1970), 23-42.

5. A. Morimoto, Stochastic stable diffeomorphisms and Takens conjecture, preprint, 1977.

6. A. Morimoto, Stochastic stability of group automorphisms, preprint, 1977.

7. J.R. Munkres, <u>Elementary Differential Topology</u>, Princeton University Press, 1966.

8. Z. Nitecki, <u>Differentiable Dynamics</u>, M.I.T. Press, 1971.

9. Z. Nitecki and M. Shub, Filtrations, decompositions and explosions, Amer. J. Math. 97(1976), 1029-1047.

10. W. Parry, Intrinsic Markov chains, Trans. Amer. Math. Soc. 112 (1964), 55-66.

11. P. Walters, Anosov diffeomorphisms are topologically stable, Topology 9(1970), 71-78.

12. P. Walters, Ergodic Theory, Springer Lecture Notes, Vol. 458, 1975.

UNIVERSITY OF WARWICK

A REFORMULATION OF COLEMAN'S CONJECTURE CONCERNING THE LOCAL CONJUGACY OF TOPOLOGICALLY HYPERBOLIC SINGULAR POINTS

by

F. Wesley Wilson, Jr.

Suppose that $f: \mathbb{R}^n, 0 \to \mathbb{R}^n, 0$ and $g: \mathbb{R}^n, 0 \to \mathbb{R}^n, 0$ are continuous functions with the property that the differential equations $\dot{x} = f(x)$ and $\dot{x} = g(x)$ generate the respective flows $F: \mathbb{R} \times \mathbb{R}^n \to \mathbb{R}^n$ and $G: \mathbb{R} \times \mathbb{R}^n \to \mathbb{R}^n$. When is there a local conjugacy between F and G near the origin, i.e., when do there exist neighborhoods N_F and N_G of 0 and a homeomorphism $h: N_F, 0 \to N_G, 0$ with the property that it carries each directed trajectory segment of F in N_F onto a directed trajectory segment of G in N_G? C. Coleman [1] has discussed this problem at some length, and has conjectured that a certain hyperbolicity condition is sufficient. In this note, we shall examine his condition and reformulate it in a more precise way by using some of the ideas from the theory of isolated invariant sets and isolating blocks. One benefit of this approach is that it will make available the generalized Lyapunov functions which one has associated with the isolating block structure [2].

The seed of this approach came from discussions with R.W. Easton, and we wish to express our appreciation for several conversations during our reflections on this topic.

§1. Some Examples of Local Conjugacies

Historically, there have been several analytic conditions given which are sufficient for the existence of local conjugacies. Since these examples have been instrumental in the formulation of Coleman's conjecture, we feel that it is appropriate to review them briefly.

Definition 1.1. An $n \times n$ matrix A is called hyperbolic if none of the eigenvalues of A is pure imaginary. The index of A is the number of eigenvalues with negative real part.

Theorem 1.2. *Suppose that $f(x) = Ax$ and $g(x) = Bx$. If A and B are hyperbolic and have the same index, then there is a (local) conjugacy $h: \mathbb{R}^n,0 \to \mathbb{R}^n,0$ for F and G. (In this case, $N_F = \mathbb{R}^n$, $N_G = \mathbb{R}^n$ and h is actually a global conjugacy).*

Corollary 1.3. *Suppose that there is a neighborhood U of 0 for which $f(x) = Ax$ and $g(x) = Bx$ for all x in U. If A and B are hyperbolic and have the same index, then there is a local conjugacy between F and G near the origin.*

The proof of Theorem 1.2 is a standard exercise in graduate courses in the qualitative theory of ordinary differential equations. The basic idea which is used in this proof is that if coordinates are chosen so that A is a block matrix with one block having only eigenvalues with negative real part and the other block having only eigenvalues with positive real part, then the system is decoupled and the flow F is a product flow. The corollary follows by restricting the global conjugacy to a sufficiently small neighborhood N_F of 0 and by taking $N_G = h(N_F)$.

Theorem 1.4 (Hartman [3]). *Suppose that $f(x) = Ax + o(x)$ where A is hyperbolic. Then there is a local conjugacy between the flows generated by $\dot{x} = f(x)$ and by $\dot{x} = Ax$ near the origin.*

Corollary 1.5. *Suppose that $f(x) = Ax + o(x)$ and $g(x) = B(x) + o(x)$ where A and B are hyperbolic and have the same index. Then there is a local conjugacy between the generated flows F and G near the origin.*

Theorem 1.6 (Coleman [4]). If F and G are asymptotically stable at the origin and if F and G have Lyapunov functions which have homeomorphic level surfaces, then there is a local conjugacy between F and G near the origin.

Corollary 1.7. If F and G are asymptotically stable at the origin and if $n \neq 4, 5$, then there is a local conjugacy between F and G near the origin.

According to [5], the level surfaces of a Lyapunov function near an asymptotically stable singular point are homotopy spheres, and whether or not they are necessarily topological spheres is equivalent to the Poincaré conjecture. If $n \neq 4$ or 5, then there is no $(n-1)$-dimensional counterexample to the Poincaré conjecture, and so the corollary follows from the theorem. We will want to avoid this point of difficulty when we paraphrase Coleman's Conjecture. The underlying idea in our approach is that if $\dot{x} = f(x)$ has one cross section near the origin which is a topological sphere, then every such cross section (and in particular all level surfaces of Lyapunov functions) is a topological sphere.

§2. Coleman's Statement

We shall recall the conjecture as it was first stated by C. Coleman [1; pages 223-224]. Part of his hypotheses were specified in the text material and then the conjecture was posed. Before recalling his words, we should note that he has labeled F and G as systems 1 and 2, respectively and that a C^0-equivalence is what we have called a local conjugacy. The words "isolated," "ball," and "hyperbolic" are not defined in [1].

> "It is assumed that the origin is an isolated stationary point and that N_F and N_G are sufficiently small balls about the origin. C^0-equivalence is always with respect to N_F and N_G."

"Conjecture: Systems 1 and 2 are C^0-equivalent if each system has complementary manifolds of stability and instability at the origin, the dimensions of the two manifolds of stability are equal, and near the origin all solutions not on these manifolds are hyperbolic."

Our interest in Coleman's conjecture stems from the fact that if it were true, then it would be a topological version of Hartman's Theorem and if it were false, then it would provide important insight into the structure of a flow on the interior of an isolating block. Unfortunately, the use of undefined terms has left the meaning of this conjecture somewhat vague. Since one is free to specify N_F and N_G as part of a partial positive result, this vagueness has caused no inconvenience in such quests. On the other hand, when one seeks to verify a counterexample, it is necessary to know precisely what it is which one is trying to contradict.

There is no reference or use of the technical notion of an "isolated invariant set" in [1]; so it is probably correct to assume that "isolated stationary point" means that there is a neighborhood of the origin which contains no other stationary points. The "hyperbolic" trajectories must certainly be "transient," i.e., they must leave N_F in finite positive and negative time. Should such a trajectory be allowed to become internally tangent to ∂N_F? The answer to this question will influence the restrictions which one imposes on the choice of the neighborhood N_F. Coleman requires that N_F be a "ball." His intent cannot be that it is a "round ball" since in that case Example 2.1 would be a trivial counterexample to Theorems 1.2, 1.3, 1.5, and 1.6. But if homeomorphic images of balls are allowed, then we are left with having to be careful with our choice of the "hyperbolicity" question.

<u>Example 2.1</u>. The following linear systems are asymptotically stable, but F has trajectories which are always transverse to the boundaries

of round balls, while G has trajectories which are externally tangent to every round ball.

$$F: \begin{array}{l} \dot{x} = -x \\ \dot{y} = -y \end{array} \qquad G: \begin{array}{l} \dot{x} = -x - 3y \\ \dot{y} = \frac{1}{3}x - y \end{array}$$

We should also observe, that if the Poincaré conjecture should turn out to be false, then the local conjugacies which are described in Theorem 1.6 may involve neighborhoods N_F and N_G which are not even topological balls. However, in our formulation, we shall specifically not allow this circumstance to enter into consideration.

§3. Isolated Invariant Sets (cf. [6], [7])

Let F be a flow on an n-manifold M and let S be a compact invariant set for F. S is said to be an <u>isolated invariant set</u> if there is a neighborhood U of S with the property that S is the largest closed invariant set in U. In this case, U is called an <u>isolating neighborhood</u> for S. In the original paper on this subject [6] Conley and Easton showed that isolated invariant sets have specially structured isolating neighborhoods which they called <u>isolating blocks</u>. The distinguishing feature of an isolating block is that if a trajectory through an interior point of the block ever reaches the boundary of the block (in positive time or in negative time) then that trajectory must leave the block before it can reenter the interior of the block in that time sense.

In [2], the author and J. Yorke provided a different proof of the existence of isolating blocks and related this structure to special kinds of generalized Lyapunov functions. Since these latter results are more appropriate to our current needs, we shall review them here.

The basic requirement is that an <u>isolating block</u> B should be an n-dimensional submanifold-with-boundary in M, i.e., B is the union of an open subset of M and its boundary ∂B, which is a topological $(n-1)$-dimensional submanifold of M. There is a decomposition of ∂B into subsets b_+, b_-, τ where $\partial B = b_+ \cup b_-$, $\tau = b_+ \cap b_-$ and b_+, b_- are topological $(n-1)$-dimensional submanifolds-with-boundary in ∂B with the property that the ingress set for F and B is $b_+ - \tau$ and the egress set for F and B is $b_- - \tau$. Thus τ is the set of points where F is tangent to B from the outside in the sense that any trajectory through a point in τ exits from B in positive time and in negative time before (possibly) entering the interior of B. It is also useful to label $A_\pm = \{x \in B \mid F(\pm t,x) \in B$ for all $t \geq 0\}$. Then $S = A_+ \cap A_-$ and $a_\pm = A_\pm \cap b_\pm$.

<u>Definition 3.1</u>. A C^r <u>isolating-block-with-corners</u> $(r > 0)$ is an isolating block B such that ∂B is a piecewise C^r submanifold of M. Specifically, we require that there be open C^r $(n-1)$-dimensional submanifolds n_+ and n_- which are transverse to F and which have the properties that $b_\pm$ is a C^r submanifold with boundary in $n_\pm$ and that $\tau = b_+ \cap b_- = n_+ \cap n_-$ is a C^r submanifold of M.

<u>Standard Example 3.2</u>. Let $F_{m,n}$ be the flow on $\mathbb{R}^{m+n}$ which is generated by

$$\dot{x} = -x$$
$$\dot{y} = y$$

where $x \in \mathbb{R}^m$, $y \in \mathbb{R}^n$. The stable manifold for this flow is $\mathbb{R}^m \times 0$ and the unstable manifold is $0 \times \mathbb{R}^n$. If D^k_ε denotes the closed ε-disk in $\mathbb{R}^k$, then $B = D^m_\varepsilon \times D^n_\varepsilon$ is a C^∞ isolating-block-with-corners for $F_{m,n}$. In this case $b_+ = \partial D^m_\varepsilon \times D^n_\varepsilon$, $b_- = D^m_\varepsilon \times \partial D^n_\varepsilon$ and $\tau = \partial D^m_\varepsilon \times \partial D^n_\varepsilon$. Therefore $A_- = 0 \times D^n_\varepsilon$, $A_+ = D^m_\varepsilon \times 0$, $a_- = 0 \times \partial D^n_\varepsilon$ and $a_+ = \partial D^m \times 0$.

Theorem 3.3. <u>If</u> B <u>is a</u> C^r <u>isolating-block-with-corners for the flow</u> F <u>and if</u> S <u>is the isolated invariant set in</u> B <u>such that</u>

$b_+ - a_+ \cong \tau \times [0,1)$ and $b_- - a_- \cong \tau \times [0,1)$, then there is a continuous function $V: B - \tau \to [-1,1]$ satisfying

1. $V(S) = 0$,

2. $\dot{V}(x) = \frac{d}{dt} V \circ F(t,x)\Big|_{t=0} < 0$ if $x \notin S$,

3. $cl(V^{-1}(0)) = V^{-1}(0) \cup \tau$ where cl denotes the closure,

4. $\{V^{-1}(c) \cap b_+ \mid c \geq 0\}$ and $\{V^{-1}(c) \cap b_- \mid c \leq 0\}$ are foliations of $b_+ - a_+$ and $b_- - a_-$, respectively.

5. $V|B - (S \cup \tau)$ is a C^∞ function.

Remark. The function V is a special form of monotone Lyapunov function for F (cf. [2: 1.4]). By using [2: 2.6] we obtain a function $U: B - \partial B \to (-1,1)$ which satisfies conditions 1, 2, 5. The modification of U to obtain V is straightforward, but technical.

§4. Coleman's Conjecture

As in the Standard Example 3.2, we shall assume that F is flow on $\mathbb{R}^m \times \mathbb{R}^n$.

Coleman's Conjecture. Let F be a flow on $\mathbb{R}^m \times \mathbb{R}^n$ which satisfies

1. The origin is an isolated invariant set for F,

2. For some $\varepsilon > 0$, $D^m_\varepsilon \times D^n_\varepsilon$ is an isolating block for 0 with $b_+ = \partial D^m_\varepsilon \times D^n_\varepsilon$, $b_- = D^m_\varepsilon \times \partial D^n_\varepsilon$, $A_+ = D^m_\varepsilon \times 0$, and $A_- = 0 \times D^n_\varepsilon$.

Then there is a homeomorphism $h: D^m_\varepsilon \times D^n_\varepsilon \to D^m_\varepsilon \times D^n_\varepsilon$ which carries the trajectory segments of F in $D^m_\varepsilon \times D^n_\varepsilon$ onto the trajectory segments of the standard example $F_{m,n}$ in $D^m_\varepsilon \times D^n_\varepsilon$.

First of all, we observe that since $a_\pm$ is transverse to the flow and $a_\pm = \partial A_\pm$ are spheres, then the level surfaces of Lyapunov functions on the stable and unstable manifolds are spheres, and so there is no possibility of having to concern ourselves with questions relating to the Poincaré Conjecture. Secondly, there are hyperbolic linear systems which do not satisfy the hypothesis that $D_\varepsilon^m \times D_\varepsilon^n$ is an isolating block (round balls don't always work, as we observed in Section 2). However, such systems do have isolating-blocks-with-corners which are homeomorphic to $D_\varepsilon^m \times D_\varepsilon^n$. Our viewpoint has been to first execute a global homeomorphism (conjugacy) of F which carries the desired isolating block and invariant manifolds onto the sets specified in our hypotheses, and then consider whether or not there exists a conjugacy onto a standard example. This viewpoint provides a better setting for the study of the validity of the conjecture.

Another benefit of this approach is that under the current hypotheses, we are assured of the existence of a monotone Lyapunov function for F in $B = D_\varepsilon^m \times D_\varepsilon^n$. The level surfaces of this function provide a foliation of codimension one for $B - (\tau \cup 0)$ whose leaves are transverse to the flow, i.e., these leaves virtually comb the trajectories as they pass through B, assuring the knottings or linkings of the trajectories are not at the root of the failure of the conjecture.

REFERENCES

1. C. Coleman, Hyperbolic stationary points, Reports of the Fifth International Conference on Nonlinear Oscillations, Vol. 2 (Qualitative Methods), Kiev (1970), 222-226.

2. F.W. Wilson and J.A. Yorke, Lyapunov functions and isolating blocks, Journal Diff. Eq. 13(1973), 106-123.

3. P. Hartman, A lemma in the theory of structural stability of differential equations, Proc. Amer. Math. Soc. 11(1960), 610-620.

4. C. Coleman, Local trajectory equivalence of differential systems, Proc. Amer. Math. Soc. 16(1965), 890-892.

5. F.W. Wilson, On the structure of the level surfaces of Lyapunov functions, Journal Diff. Eq. 3(1967), 323-329.

6. C. Conley and R. Easton, Isolated invariant sets and isolating blocks, Trans. Amer. Math. Soc. 158(1971), 1-27.

UNIVERSITY OF COLORADO

ERGODIC ACTIONS AND STOCHASTIC PROCESSES ON GROUPS AND HOMOGENEOUS SPACES

by

Robert J. Zimmer

§1. Introduction

Skew products have received considerable attention in ergodic theory both as an important source of examples and as an object of study in their own right. In this paper we shall see how two basic, seemingly unrelated, types of problems, one concerning the restriction of ergodic actions of a group to a closed subgroup and the other concerning stochastic processes on groups and homogeneous spaces, can both be converted to questions concerning skew products, and can then both be solved, at least in many cases, by a unified technique. Some of the results in this paper have or will appear elsewhere, and we shall not reproduce these proofs here. However, new examples and results are given in reasonably full detail. An interesting feature of some of the proofs is that they provide a new mode of application of results from the theory of unitary representations of groups to ergodic theory and probability.

We begin by recalling the definition of skew products. We shall actually for the most part restrict attention to a special class of skew products, namely those in which the fiber is a locally compact group H, or more generally, a homogeneous space H/K, where $K \subset H$ is a closed subgroup. Let (S,m) be a standard Borel space with a probability measure. Suppose $T: S \to S$ is an invertible and measurable transformation, and that T leaves m quasi-invariant (i.e., $m(TA) = 0$ if and only if $m(A) = 0$ for $A \subset S$ measurable), and T is ergodic. We shall often want m to actually be invariant, i.e., $m(TA) = m(A)$, and we shall be explicit when we are using this stronger condition. Suppose $a: S \to H$ is a Borel function, where H is a locally compact separable group. Then we define a skew product transformation $\tilde{T}$ on $S \times H/K$, where K is a closed (possibly trivial) subgroup of H, by $\tilde{T}(s,x) = (Ts, xa(s))$, where $xa(s)$ refers to the natural right action of H on H/K. There is a unique measure class on H/K quasi-invariant under H, and the product measure on $S \times H/K$ will then be quasi-invariant under $\tilde{T}$. Of course if m is invariant and there is a finite H-invariant measure on H/K, then $\tilde{T}$ will actually leave the product measure invariant. When endowed with this action, we shall denote the product space by $S \times_a H/K$.

One can define skew products for actions of more general groups than the integers. Thus, suppose G is a locally compact separable group which acts on the right on S so that each $g \in G$ leaves m quasi-invariant and the G-action is ergodic on S. Let $a: S \times G \to H$ be a Borel function. Then we define a skew product action of G on $S \times H/K$ by $(s,x)g = (sg, xa(s,g))$. In order for this to actually define an action, a must satisfy a certain condition, namely it must be a cocycle, i.e., $a(s,gh) = a(s,g)a(sg,h)$. Once again, we denote the skew product G-space by $S \times_a H/K$

The general problem concerning skew products that we will deal with is to determine when the skew product transformation or, more generally, action is or is not ergodic.

§2. The Problems

In this section we shall describe the problems alluded to in the introduction and see how they can be reduced to the question of ergodicity of skew products.

A. Restricting ergodic actions to subgroups

Suppose a group G acts ergodically on S, and that Γ is a closed subgroup of G. We wish to determine when Γ will also be ergodic on S. The reduction to a skew product is accomplished by the following fact.

Proposition 2.1 [18, Theorem 4.2]. Γ <u>is ergodic on</u> S <u>if and only if</u> G <u>is ergodic on</u> $S \times G/\Gamma$.

Thus, the ergodicity of Γ is reduced to the ergodicity of a product action, which is of course a special case of a skew product, obtained here by taking $a: S \times G \to G$ to be the cocycle $a(s,g) = g$.

B. Random processes

Suppose $Z_1, Z_2, \ldots$ is a sequence of independent identically distributed random variables taking values in a group H. Let the common distribution of the Z_n be μ, a probability measure on H. For each $h \in H$, we can then define the random walk on H with law μ starting at h to be the sequence of random variables $X_0 = h, \ldots, X_n = hZ_1Z_2\ldots Z_n$. Similarly, we can define a random walk on H/K starting at $x \in H/K$ by $X_0 = x, \ldots, X_n = xZ_1\ldots Z_n$. The Z_n are called the increments of the process. The random walk is called recurrent if for each $A \subset H/K$ of positive measure, $P(X_n \in A$ for infinitely many $n \mid X_0 = x) = 1$ for almost all $x \in H/K$. (We remark that this is weaker than the notion of Harris recurrence which requires that this condition hold for all x. In the case of groups, though not for homogeneous spaces, this notion of recurrence is equivalent to topological recurrence [9, Exercise 4.22].) Typical questions that arise are to determine which groups H admit recurrent random walks and to determine when a measure μ on H defines a recurrent random walk on H or H/K. These problems have received very considerable attention. It has been known for some time, for example, that Z (the integers) and Z^2 admit recurrent random walks, while Z^n does not for $n \geq 3$.

A related notion to recurrence is ergodicity. Haar measure on H, or an H-invariant measure on H/K, if it exists, will be an invariant measure ν for the random walk, and one can then, using ν as an initial distribution, extend X_n to a 2-sided random walk, so that X_n is defined for all $n \in Z$. One can then form the space of sample seqeunces for the 2-sided random walk. (See [3] or [10] for example.) The random walk is called ergodic if the shift on the 2-sided sample sequence space is ergodic. One then has the basic result that recurrence and ergodicity are equivalent for random walks [3], a fact which depends on the Markovian character of the sequence X_n.

A natural generalization of this problem is to study the situation in which the assumption of the independence of the increments is relaxed. Thus suppose Z_n is a stationary, ergodic, H-valued stochastic process. Form X_n as above. One can now inquire as to the existence of processes on groups with stationary, ergodic increments that are recurrent or ergodic. We can also ask about the situation in which the increments are required to satisfy some condition weaker than independence, but stronger than stationary and ergodic -- for example, ergodic and finite valued, purely non-deterministic, etc. It is important to observe that once the independence assumptions on Z_n are weakened, X_n will no longer necessarily be Markovian and equivalence of ergodicity and recurrence no longer holds.

All of these problems can be reformulated in terms of skew products as follows. Let I be a standard Borel space and $\Omega = \prod_{-\infty}^{\infty} I$. Let $T: \Omega \to \Omega$ be the shift, i.e., $(T\omega)(n) = \omega(n+1)$. Let m be a probability measure on Ω which is invariant and ergodic under T. Let H be a group, K a closed subgroup, and $f: I \to H$. Let $a: \Omega \to H$ be defined by $a(\omega) = f(\omega(0))$. Let $\tilde{T}$ be the skew product transformation on $\Omega \times_a H/K$.

Proposition 2.2. $\tilde{T}$ _is isomorphic to the shift on the sample sequence space of the 2-sided stochastic process with stationary ergodic increments_ Z_n, _where_ Z_n _is defined by_ $Z_n(\omega) = \omega(n)$.

For the proof, see [14, Theorem 3], for example. Thus,

ergodicity of the process is equivalent to ergodicity of the skew product.

The results we spell out in the following sections concentrate on the condition of ergodicity, obtaining results about recurrence when the two conditions are equivalent. For results about recurrence in the general case, focusing on some different questions, the reader is referred to [11].

§3. Groups Admitting Ergodic Processes With Stationary Purely Non-Deterministic Finite-Valued Increments

In this section, we provide some partial answers to one of the questions raised in the previous paragraph. Namely, we identify certain groups which admit ergodic processes whose increments are stationary, finite-valued, and purely non-deterministic [10]. Processes with stationary finite-valued independent increments are just random walks whose law has finite support. Thus, in the independent case, the only free abelian groups admitting such an ergodic process are Z and Z^2. The next result shows that relaxing the condition of independence to the condition of purely non-deterministic admits many more groups. We remark that in terms of skew products, finite-valued independent increments corresponds to $T: \Omega \to \Omega$ being a Bernoulli shift on a finite state space, and finite-valued purely non-deterministic increments corresponds to T being a K-shift on a finite state space.

Theorem 3.1. <u>Let H be the product of a finitely generated discrete nilpotent group with a connected nilpotent Lie group. Then there is a K-shift</u> $T: \Omega \to \Omega$, $\Omega = \prod_{-\infty}^{\infty} I$, I <u>a finite set, and a function</u> $f: I \to H$ <u>such that the skew product</u> $\tilde{T}$ <u>on</u> $\Omega \times_a H$ <u>is ergodic, where</u> $a(\omega) = f(\omega(0))$. <u>In other words, H admits an ergodic process with stationary purely non-deterministic finite-valued increments.</u>

Proof: The proof consists of examining in greater detail a construction given in [15], and applying a theorem of Meilijson which gives a criterion for a transformation to be a K-shift. Namely, suppose G is a locally compact second countable group and $H \subset G$ is a closed central subgroup. Suppose that both H and G/H admit ergodic processes with stationary, finite-valued, purely non-deterministic increments. We will show that G must then admit such a process. The same technique shows that the product of two groups admitting such a process must itself admit such a process. The integers and the real line admit such processes (which are in fact random walks), and the theorem will then follow by induction, first to finitely generated abelian groups and connected abelian Lie groups, then finally to the nilpotent case. Here one uses the fact that a subgroup of a finitely generated nilpotent group is also finitely generated. We now proceed to the inductive step.

Suppose for $i = 1,2$, $\Omega_i = \prod_{-\infty}^{\infty} I_i$, where I_i is a finite set, and μ_i is a shift invariant probability measure on Ω_i such that the shift T_i on Ω_i is a K-shift. Suppose $f_1: I_1 \to H$ with $\Omega_1 \times_{\tilde{f}_1} H$ ergodic, where $\tilde{f}_1(\omega) = f_1(\omega(0))$, and $f_2: I_2 \to G/H$ such that $\Omega_2 \times_{\tilde{f}_2} G/H$ is ergodic, where $\tilde{f}_2(\omega) = f_2(\omega(0))$. Choose an arbitrary finite-valued Borel lift of the function $\tilde{f}_2$ to a function $g_2: \Omega_2 \to G$. Via the product action, $\Omega_1 \times \Omega_2$ becomes a Z^2-space, and since H is in the center of G, there is a cocycle $\beta: \Omega_1 \times \Omega_2 \times Z^2 \to G$ such that $\beta(\omega_1,\omega_2,(1,0)) = \tilde{f}_1(\omega_1)$ and $\beta(\omega_1,\omega_2,(0,1)) = g_2(\omega_2)$. Thus, the cocycle β has the property that for each $(n,m) \in Z^2$, $\beta(\omega_1,\omega_2,(n,m))$ takes on only finitely many values. Now let Ω be a Bernouli shift on a finite state space I, and let $\alpha: I \to Z^2$ be a finite-valued function defining a recurrent random walk. We form the skew product space $S = \Omega \times_\alpha (\Omega_1 \times \Omega_2)$, and let $f: S \to G$ be defined by $f(\omega,\omega_1,\omega_2) = \beta(\omega_1,\omega_2,\alpha(\omega))$. An examination of the proofs of [15, Lemma 3.2 and Theorem 3.6] shows that $S \times_f G$ is actually ergodic. Furthermore, f is finite-valued because of the corresponding properties of α and β. To see that the skew products on S is actually a K-shift, note that $\Omega \times_\alpha (\Omega_1 \times \Omega_2)$ is isomorphic to the fibered product [13] $(\Omega \times_{\alpha_1} \Omega_1) \times_\Omega (\Omega \times_{\alpha_2} \Omega_2)$, where α_1 and α_2 are the coordinate functions of α. Since α defines a recurrent random walk on Z^2, α_1 and α_2 define recurrent random walks on Z, and it follows from a theorem of Meilijson [5] that $\Omega \times_{\alpha_i} \Omega_i$ are both K-automorphisms. The proof that the product of K-automorphisms is again a K-automorphism can be carried over to fibered products, and we conclude that S is a K-automorphism. Since f is finite-valued, S will have a factor space S_0 which can be represented as a finite state K-shift, and f will then be measurable with respect to the state space. Then $f: S_0 \to G$ is the required function.

We remark that since any compact group possessing a finite set which generates a dense subgroup admits a recurrent random walk with finite-valued increments, the technique of the above proof shows that Theorem 3.1 remains true if H is a product of a finitely generated nilpotent discrete group, a connected nilpotent Lie group, and such a compact group. In a converse direction, we have the following.

Theorem 3.2 [17, Theorem 3.1]. <u>If H admits an ergodic process with stationary increments, then H is amenable.</u>

Further results on this question can be found in [12], [15]. It would be interesting to determine whether or not Theorem 3.2 is still true if ergodicity is replaced by recurrence.

§4. Results on Homogeneous Spaces

In this section, we present results on skew products whose fibers are homogeneous spaces with finite invariant measure, and thus provide answers in certain cases to the questions concerning random processes and restrictions of ergodic actions raised in Section 2. The proofs require a significant amount of material from the theory of unitary representations. The results stated below for homogeneous spaces of nilpotent and semisimple Lie groups can be found in [16] and [18]. Here, in Theorem 4.7 below, we shall illustrate the method of proof in the nilpotent case by showing how the technique can be extended to apply to a class of solvable, non-nilpotent semi-direct products.

Theorem 4.1 [16]. *Let* N *be a connected simply connected nilpotent Lie group and let* Γ *be a lattice subgroup, i.e.,* Γ *is discrete and* N/Γ *has a finite* N*-invariant measure. Then if* G *is any locally compact separable group acting ergodically on a space* S, *and* $a: S \times G \to N$ *is a cocycle, then the skew product* $S \times_a N/\Gamma$ *is ergodic if and only if* $S \times_a N/\Gamma[N,N]$ *is ergodic.*

Corollary 4.2. *If* N *acts ergodically on a space* S, *then* Γ *is ergodic on* S *if and only if* $\Gamma[N,N]$ *is ergodic on* S. *In particular, if* $[N,N]$ *is ergodic,* Γ *is ergodic for every lattice* Γ.

Corollary 4.3. *A stochastic process on* N *with stationary ergodic increments induces an ergodic process on* N/Γ *if and only if it induces an ergodic process on* $N/\Gamma[N,N]$. *In particular, a random walk on* N/Γ *is recurrent if and only if the random walk on* $N/\Gamma[N,N]$ *is recurrent.*

Theorem 4.4 [18]. *Let* H *be a connected, non-compact, simple Lie group with finite center, and let* $\Gamma \subset H$ *be a lattice subgroup. If* G *is ergodic on* S *and* $a: S \times G \to H$ *is a cocycle, then* $S \times_a H/\Gamma$ *is ergodic if and only if* a *is not cohomologous to a cocycle into a compact subgroup of* H, *that is, if and only if there does not exist a function* $f: S \to H$ *and a compact subgroup* K *such that for each*

$g \in G$, $f(s)a(s,g)f(sg)^{-1} \in K$ a.e.

Corollary 4.5 [18]. If X is a properly ergodic H-space (i.e., every orbit is a null set), then Γ is also ergodic on X.

To deduce Corollary 4.5 from Theorem 4.4 requires some work. See the techniques of [18] for the proof.

Corollary 4.6. If μ is a measure on H, then μ induces a recurrent random walk on H/Γ if and only if μ is not supported by a compact subgroup of H.

While Corollary 4.6 can presumably be obtained by more standard methods by using the results of Moore [6], it is perhaps of interest to see how it can be deduced directly from Theorem 4.4.

Proof (of Corollary 4.6): Let $\Omega = \prod_{-\infty}^{\infty} (I,\nu)$, T the (generalized Bernoulli) shift on Ω, and $a: I \to H$ with $a_*(\nu) = \mu$. Let $\alpha: \Omega \times Z \to H$ be the cocycle corresponding to the function a. If the support of μ is contained in a compact subgroup K, then α takes values in K almost surely, and so the skew product $\Omega \times_\alpha H/\Gamma$ is not ergodic, and hence the random walk is not recurrent. Conversely, let L be the closed subgroup generated by the support of μ and assume L is not compact. Suppose first that the random walk on L with law μ is topologically recurrent. Then $\Omega \times_\alpha L$ is ergodic. This implies that the range-closure of the cocycle $\alpha: \Omega \times Z \to H$ is the H-space H/L [8]. By [18, Theorem 6.1], the skew product transformation $\Omega \times_\alpha H/\Gamma$ will be ergodic if the product H-space $H/L \times H/\Gamma$ is ergodic, which is in turn equivalent to the ergodicity of L on H/Γ. But since L is not compact, this ergodicity follows from [6].

On the other hand, consider the case in which the random walk on L is not topologically recurrent but transient. Then examination of the proof of [14, Theorem 3] shows that for almost all ω, $\alpha(\omega,n) \to \infty$ (in that it eventually leaves every compact subset) as $n \to \infty$.

Suppose α is cohomologous to a cocycle into K, where K is a compact subgroup, i.e., there is a function $\varphi: \Omega \to H$ such that $\varphi(\omega)\alpha(\omega,n)\varphi(\omega\cdot n)^{-1} = \beta(\omega,n) \in K$ for each n and almost all ω. Here $\omega\cdot n$ denotes the action of Z on Ω by the shift. By ergodicity of the shift on Ω, for almost all ω, there is a sequence $n_k \to \infty$ such that $\varphi(\omega\cdot n_k)$ are all contained in a compact set. It follows that $\alpha(\omega,n_k) = \varphi(\omega)^{-1}\beta(\omega,n_k)\varphi(\omega\cdot n_k)$ cannot go to infinity, which is a contradiction. The corollary now follows from Theorem 4.4.

We remark that Theorem 4.4 and its corollaries generalize to the case in which G is a product of simple Lie groups and Γ is an irreducible lattice in G, and to simple algebraic groups over more general fields [18].

We now examine a class of solvable semi-direct product Lie groups. The proof in this case is essentially an extension of the proof in the nilpotent case, but is at points made somewhat easier by the realization of the group as a semi-direct product. For each $n \times n$ matrix A, we define a semi-direct product group $S_A = R^n \circledS R$, where R^n is normal in S_A and the action of R on R^n is given by the matrix $\exp(tA)$. Then S_A is a solvable Lie group, and will be nilpotent if and only if the matrix A is nilpotent. We suppose that A has no purely imaginary eigenvalues, and in that case, S_A is an exponential Lie group, that is, the exponential map on the Lie algebra of S_A is bijective. As with nilpotent Lie groups, a group of the form S_A may or may not admit a lattice subgroup. See [1, Theorem III.3,2] for example.

Theorem 4.7. <u>Suppose</u> Γ <u>is a lattice subgroup of the solvable exponential group</u> S_A. <u>If</u> $a: S \times G \to S_A$ <u>is a cocycle, then</u> $S \times_a S_A/\Gamma$ <u>is ergodic if and only if</u> $S \times_a /\Gamma[S_A,S_A]$ <u>is ergodic</u>.

(We note that $\Gamma[S_A,S_A]$ is a closed subgroup [1, Chapter VII] and that $S_A/\Gamma[S_A,S_A]$ is a torus.)

Corollary 4.8. __If an ergodic action of S_A is still ergodic when restricted to $[S_A,S_A]$, it is ergodic when restricted to any lattice subgroup of S_A.__

__Proof__ (of Theorem 4.7): As the nilpotent case follows from [16], we shall assume S_A is non-nilpotent. If $S \times {}_a S_A/\Gamma$ is ergodic, clearly $S \times {}_a S_A/\Gamma[S_A,S_A]$ is also, and so it suffices to prove the converse. Let π be the representation of S_A induced by the identity representation of Γ, i.e., π is a translation in $L^2(S_A/\Gamma)$. Then $\pi = \sum^{\oplus} \sigma_i \oplus \sum^{\oplus} \pi_i$ where σ_i are one dimensional and π_i are infinite dimensional irreducible representations of S_A. One can now apply Mackey's theory for semi-direct products to determine the representations π_i. The normal abelian subgroup R^n is regularly embedded [2, Cor. I.3.10], and so one obtains all irreducible representations by examining the orbits of S_A in $\hat{R}^n$ [4]. One readily deduces that each π_i is induced by a one dimensional representation θ_i of R^n. At this point, we have need of the theory of unitary cocycles of ergodic actions, for which the reader is referred to [13] and [18]. Suppose $S \times {}_a S_A/\Gamma[S_A,S_A]$ is ergodic but that $S \times {}_a S_A/\Gamma$ is not. Then for some i, the unitary cocycle $\pi_i \circ a$ must contain the identity cocycle [18, Proposition 3.1]. Arguing as in [16, Theorem 3.2], this implies that $U \circ a$ must contain the identity cocycle, where U is the representation of S_A induced by the identity representation of R^n. Since the space of orbits in $L^2(R)$ under the representation U form a standard Borel space [16, Corollary 2.8], arguing as in [16], we conclude that a must be equivalent to a cocycle into R^n. To complete the proof, it thus suffices to see that R^n is not ergodic on $S_A/\Gamma[S_A,S_A]$, for this will contradict the ergodicity of $S \times {}_a S/\Gamma[S_A,S_A]$. But R^n is the maximal connected closed nilpotent subgroup of S_A, and hence $R^n\Gamma$ is closed in S_A [7, Cor. III.3.5]. Since $R^n \supset [S_A,S_A]$, the action of R^n is clearly not minimal on the torus $S_A/\Gamma[S_A,S_A]$, and hence not ergodic.

REFERENCES

1. L. Auslander, L. Green and F. Hahn, Flows on homogeneous spaces, Annals of Math. Studies, no. 53, Princeton, 1963.

2. P. Bernat et. al., Representations des Groupes de Lie Resoluble, Dunod, Paris, 1972.

3. T.E. Harris and H. Robbins, Ergodic theory of Markov chains admitting an infinite invariant measure, Proc. Nat. Acad. Sci. U.S.A., 39(1953), 860-864.

4. G.W. Mackey, Induced representations of locally compact groups, I, Annals of Math. 55(1952), 101-139.

5. I. Meilijson, Mixing properties of a class of skew products, Israel J. Math., 19(1974), 266-270.

6. C.C. Moore, Ergodicity of flows on homogeneous spaces, Amer. J. Math. 88(1966), 154-178.

7. M.S. Raghunathan, Discrete Subgroups of Lie Groups, Springer-Verlag, New York, 1972.

8. A. Ramsay, Virtual groups and group actions, Advances in Math., 6(1971), 253-322.

9. D. Revuz, Markov Chains, North-Holland, Amsterdam, 1975.

10. M. Rosenblatt, Markov Process. Structure and Asymptotic Behavior, Springer-Verlag, New York, 1971.

11. K. Schmidt, Lectures on cocycles of ergodic transformation groups, preprint.

12. J.J. Westman, Virtual group homomorphisms with dense range, Ill. J. of Math. 20(1976), 41-47.

13. R.J. Zimmer, Extensions of ergodic group actions, Ill. J. of Math. 20(1976), 373-409.

14. R.J. Zimmer, Random walks on compact groups and the existence of cocycles, Israel J. Math., 26(1977), 84-90.

15. R.J. Zimmer, Cocycles and the structure of ergodic group actions, Israel J. Math. 26(1977), 214-220.

16. R.J. Zimmer, Compact nilmanifold extensions of ergodic actions, Trans. Amer. Math. Soc. 223(1976), 397-406.

17. R.J. Zimmer, Amenable ergodic group actions and an application to Poisson boundaries of random walks, to appear, J. Funct. Anal.

18. R.J. Zimmer, Orbit spaces of unitary representations, ergodic theory, and simple Lie groups, to appear, Annals of Math.

UNIVERSITY OF CHICAGO

Vol. 489: J. Bair and R. Fourneau, Etude Géométrique des Espaces Vectoriels Une Introduction. VII, 185 pages. 1975.

Vol. 490: The Geometry of Metric and Linear Spaces. Proceedings 1974 Edited by L M Kelly X, 244 pages 1975.

Vol. 491: K. A. Broughan, Invariants for Real-Generated Uniform Topological and Algebraic Categories. X, 197 pages. 1975.

Vol. 492: Infinitary Logic: In Memoriam Carol Karp. Edited by D W. Kueker. VI, 206 pages. 1975.

Vol. 493: F. W. Kamber and P. Tondeur, Foliated Bundles and Characteristic Classes. XIII, 208 pages. 1975.

Vol. 494: A Cornea and G. Licea. Order and Potential Resolvent Families of Kernels. IV, 154 pages 1975.

Vol. 495. A. Kerber, Representations of Permutation Groups II. V, 175 pages. 1975.

Vol. 496: L. H. Hodgkin and V. P. Snaith, Topics in K-Theory. Two Independent Contributions. III, 294 pages. 1975.

Vol. 497: Analyse Harmonique sur les Groupes de Lie. Proceedings 1973-75. Edité par P. Eymard et al. VI, 710 pages. 1975.

Vol. 498: Model Theory and Algebra. A Memorial Tribute to Abraham Robinson. Edited by D. H. Saracino and V. B. Weispfenning. X, 463 pages. 1975.

Vol. 499: Logic Conference, Kiel 1974 Proceedings. Edited by G. H. Müller, A. Oberschelp, and K. Potthoff. V, 651 pages 1975.

Vol. 500: Proof Theory Symposion, Kiel 1974. Proceedings Edited by J. Diller and G. H. Müller. VIII, 383 pages. 1975.

Vol. 501: Spline Functions, Karlsruhe 1975. Proceedings. Edited by K. Böhmer, G. Meinardus, and W. Schempp. VI, 421 pages. 1976.

Vol. 502: János Galambos, Representations of Real Numbers by Infinite Series. VI, 146 pages. 1976.

Vol. 503: Applications of Methods of Functional Analysis to Problems in Mechanics Proceedings 1975. Edited by P. Germain and B. Nayroles. XIX, 531 pages 1976.

Vol. 504. S. Lang and H. F. Trotter, Frobenius Distributions in GL_2-Extensions. III, 274 pages. 1976

Vol. 505: Advances in Complex Function Theory. Proceedings 1973/74. Edited by W. E. Kirwan and L. Zalcman. VIII, 203 pages. 1976.

Vol. 506: Numerical Analysis, Dundee 1975 Proceedings. Edited by G A. Watson X, 201 pages. 1976.

Vol 507: M C. Reed, Abstract Non-Linear Wave Equations. VI, 128 pages 1976.

Vol. 508: E Seneta, Regularly Varying Functions. V, 112 pages 1976

Vol 509: D E Blair, Contact Manifolds in Riemannian Geometry. VI, 146 pages 1976

Vol. 510: V. Poènaru, Singularités C^∞ en Présence de Symétrie. V, 174 pages. 1976.

Vol. 511: Séminaire de Probabilités X. Proceedings 1974/75. Edite par P. A. Meyer. VI, 593 pages. 1976.

Vol. 512: Spaces of Analytic Functions, Kristiansand, Norway 1975. Proceedings. Edited by O. B. Bekken, B. K. Øksendal, and A. Stray VIII, 204 pages. 1976.

Vol. 513: R. B. Warfield, Jr. Nilpotent Groups. VIII, 115 pages 1976.

Vol. 514: Séminaire Bourbaki vol. 1974/75. Exposés 453 - 470. IV, 276 pages. 1976.

Vol. 515: Bäcklund Transformations. Nashville, Tennessee 1974. Proceedings. Edited by R. M. Miura. VIII, 295 pages. 1976.

Vol 516: M. L. Silverstein, Boundary Theory for Symmetric Markov Processes. XVI, 314 pages. 1976.

Vol. 517: S. Glasner, Proximal Flows VIII, 153 pages. 1976.

Vol. 518: Séminaire de Théorie du Potentiel, Proceedings Paris 1972-1974. Edité par F. Hirsch et G Mokobodzki. VI, 275 pages. 1976.

Vol. 519: J. Schmets, Espaces de Fonctions Continues. XII, 150 pages. 1976.

Vol 520: R. H. Farrell, Techniques of Multivariate Calculation. X, 337 pages. 1976.

Vol. 521: G. Cherlin, Model Theoretic Algebra - Selected Topics. IV, 234 pages. 1976.

Vol. 522: C. O. Bloom and N. D. Kazarinoff, Short Wave Radiation Problems in Inhomogeneous Media: Asymptotic Solutions. V. 104 pages. 1976.

Vol. 523: S. A. Albeverio and R. J. Høegh-Krohn, Mathematical Theory of Feynman Path Integrals. IV, 139 pages. 1976.

Vol. 524: Séminaire Pierre Lelong (Analyse) Année 1974/75. Edité par P. Lelong. V, 222 pages. 1976.

Vol. 525: Structural Stability, the Theory of Catastrophes, and Applications in the Sciences. Proceedings 1975. Edited by P. Hilton. VI, 408 pages. 1976.

Vol. 526: Probability in Banach Spaces. Proceedings 1975. Edited by A. Beck. VI, 290 pages. 1976.

Vol. 527: M. Denker, Ch. Grillenberger, and K. Sigmund, Ergodic Theory on Compact Spaces. IV, 360 pages. 1976.

Vol. 528: J. E. Humphreys, Ordinary and Modular Representations of Chevalley Groups III, 127 pages 1976

Vol. 529: J. Grandell, Doubly Stochastic Poisson Processes. X, 234 pages. 1976.

Vol. 530: S. S. Gelbart, Weil's Representation and the Spectrum of the Metaplectic Group. VII, 140 pages. 1976.

Vol. 531: Y.-C. Wong, The Topology of Uniform Convergence on Order-Bounded Sets. VI, 163 pages. 1976.

Vol. 532: Théorie Ergodique. Proceedings 1973/1974. Edité par J.-P. Conze and M S. Keane. VIII, 227 pages. 1976.

Vol. 533: F. R Cohen, T. J. Lada, and J. P. May, The Homology of Iterated Loop Spaces. IX, 490 pages. 1976.

Vol. 534: C. Preston, Random Fields. V, 200 pages. 1976.

Vol. 535: Singularités d'Applications Differentiables. Plans-sur-Bex. 1975. Edité par O. Burlet et F. Ronga. V, 253 pages. 1976.

Vol. 536: W. M. Schmidt, Equations over Finite Fields. An Elementary Approach. IX, 267 pages. 1976.

Vol. 537: Set Theory and Hierarchy Theory. Bierutowice, Poland 1975. A Memorial Tribute to Andrzej Mostowski. Edited by W. Marek, M. Srebrny and A. Zarach. XIII, 345 pages. 1976.

Vol. 538: G. Fischer, Complex Analytic Geometry. VII, 201 pages. 1976.

Vol. 539: A. Badrikian, J. F. C. Kingman et J. Kuelbs, Ecole d'Eté de Probabilités de Saint Flour V-1975. Edité par P.-L. Hennequin. IX, 314 pages. 1976.

Vol. 540: Categorical Topology, Proceedings 1975. Edited by E. Binz and H Herrlich. XV, 719 pages. 1976.

Vol. 541: Measure Theory, Oberwolfach 1975. Proceedings. Edited by A. Bellow and D. Kölzow. XIV, 430 pages. 1976.

Vol. 542: D. A. Edwards and H. M Hastings, Čech and Steenrod Homotopy Theories with Applications to Geometric Topology. VII, 296 pages. 1976.

Vol. 543. Nonlinear Operators and the Calculus of Variations, Bruxelles 1975. Edited by J. P. Gossez, E. J. Lami Dozo, J. Mawhin, and L. Waelbroeck, VII, 237 pages. 1976.

Vol. 544: Robert P. Langlands, On the Functional Equations Satisfied by Eisenstein Series. VII, 337 pages. 1976.

Vol. 545: Noncommutative Ring Theory. Kent State 1975 Edited by J. H. Cozzens and F. L. Sandomierski. V, 212 pages. 1976.

Vol. 546: K. Mahler, Lectures on Transcendental Numbers. Edited and Completed by B. Diviš and W. J. Le Veque. XXI, 254 pages. 1976.

Vol. 547: A. Mukherjea and N. A. Tserpes, Measures on Topological Semigroups: Convolution Products and Random Walks. V, 197 pages. 1976.

Vol. 548: D. A. Hejhal, The Selberg Trace Formula for PSL (2,ℝ). Volume I. VI, 516 pages. 1976.

Vol. 549: Brauer Groups, Evanston 1975. Proceedings. Edited by D. Zelinsky. V, 187 pages. 1976.

Vol 550: Proceedings of the Third Japan - USSR Symposium on Probability Theory Edited by G. Maruyama and J. V. Prokhorov. VI, 722 pages. 1976.

Vol. 551: Algebraic K-Theory, Evanston 1976. Proceedings. Edited by M. R. Stein. XI, 409 pages. 1976.

Vol. 552: C. G. Gibson, K. Wirthmüller, A. A. du Plessis and E. J. N. Looijenga. Topological Stability of Smooth Mappings. V, 155 pages. 1976.

Vol. 553: M. Petrich, Categories of Algebraic Systems. Vector and Projective Spaces, Semigroups, Rings and Lattices. VIII, 217 pages. 1976.

Vol. 554: J. D. H. Smith, Mal'cev Varieties. VIII, 158 pages. 1976.

Vol. 555: M. Ishida, The Genus Fields of Algebraic Number Fields. VII, 116 pages. 1976.

Vol. 556: Approximation Theory. Bonn 1976. Proceedings. Edited by R. Schaback and K. Scherer VII, 466 pages. 1976.

Vol. 557: W Iberkleid and T. Petrie, Smooth S^1 Manifolds. III, 163 pages. 1976.

Vol. 558: B. Weisfeiler, On Construction and Identification of Graphs. XIV, 237 pages. 1976.

Vol. 559: J.-P. Caubet, Le Mouvement Brownien Relativiste. IX, 212 pages. 1976.

Vol. 560: Combinatorial Mathematics, IV, Proceedings 1975. Edited by L. R. A. Casse and W. D. Wallis. VII, 249 pages. 1976.

Vol. 561: Function Theoretic Methods for Partial Differential Equations. Darmstadt 1976. Proceedings. Edited by V. E. Meister, N. Weck and W. L. Wendland. XVIII, 520 pages. 1976.

Vol. 562. R. W. Goodman, Nilpotent Lie Groups: Structure and Applications to Analysis. X, 210 pages 1976.

Vol. 563: Séminaire de Théorie du Potentiel. Paris, No. 2. Proceedings 1975–1976. Edited by F. Hirsch and G. Mokobodzki. VI, 292 pages. 1976.

Vol. 564: Ordinary and Partial Differential Equations, Dundee 1976. Proceedings. Edited by W. N. Everitt and B. D. Sleeman. XVIII, 551 pages. 1976.

Vol. 565: Turbulence and Navier Stokes Equations Proceedings 1975. Edited by R. Temam. IX, 194 pages. 1976.

Vol. 566: Empirical Distributions and Processes. Oberwolfach 1976. Proceedings. Edited by P. Gaenssler and P. Révész VII, 146 pages. 1976.

Vol. 567: Séminaire Bourbaki vol. 1975/76. Exposés 471–488 IV, 303 pages. 1977.

Vol 568: R. E. Gaines and J. L. Mawhin, Coincidence Degree, and Nonlinear Differential Equations V, 262 pages. 1977.

Vol. 569: Cohomologie Etale SGA 4½. Séminaire de Géométrie Algébrique du Bois-Marie. Edité par P. Deligne. V, 312 pages. 1977.

Vol. 570: Differential Geometrical Methods in Mathematical Physics, Bonn 1975. Proceedings. Edited by K. Bleuler and A. Reetz. VIII, 576 pages. 1977.

Vol. 571: Constructive Theory of Functions of Several Variables, Oberwolfach 1976. Proceedings. Edited by W. Schempp and K. Zeller. VI. 290 pages 1977

Vol. 572: Sparse Matrix Techniques, Copenhagen 1976. Edited by V. A. Barker. V, 184 pages. 1977.

Vol. 573: Group Theory, Canberra 1975. Proceedings. Edited by R. A. Bryce, J. Cossey and M. F. Newman. VII, 146 pages. 1977.

Vol. 574: J. Moldestad, Computations in Higher Types. IV, 203 pages. 1977.

Vol. 575: K-Theory and Operator Algebras, Athens, Georgia 1975. Edited by B B. Morrel and I. M. Singer. VI, 191 pages. 1977.

Vol. 576: V. S. Varadarajan, Harmonic Analysis on Real Reductive Groups. VI, 521 pages. 1977.

Vol. 577: J. P. May, E_∞ Ring Spaces and E_∞ Ring Spectra. IV, 268 pages. 1977.

Vol. 578: Séminaire Pierre Lelong (Analyse) Année 1975/76. Edité par P. Lelong. VI, 327 pages. 1977.

Vol. 579: Combinatoire et Représentation du Groupe Symétrique, Strasbourg 1976. Proceedings 1976. Edité par D. Foata IV, 339 pages. 1977.

Vol. 580: C. Castaing and M. Valadier, Convex Analysis and Measurable Multifunctions. VIII, 278 pages. 1977.

Vol. 581: Séminaire de Probabilités XI, Université de Strasbourg. Proceedings 1975/1976. Edité par C. Dellacherie, P. A. Meyer et M. Weil. VI, 574 pages. 1977.

Vol. 582: J. M. G. Fell, Induced Representations and Banach *-Algebraic Bundles. IV, 349 pages. 1977

Vol. 583: W. Hirsch, C. C. Pugh and M. Shub, Invariant Manifolds. IV, 149 pages. 1977.

Vol 584: C. Brezinski, Accélération de la Convergence en Analyse Numérique. IV, 313 pages. 1977.

Vol. 585: T. A. Springer, Invariant Theory. VI, 112 pages. 1977.

Vol. 586: Séminaire d'Algèbre Paul Dubreil, Paris 1975–1976 (29ème Année). Edited by M. P. Malliavin. VI, 188 pages. 1977.

Vol. 587: Non-Commutative Harmonic Analysis. Proceedings 1976. Edited by J. Carmona and M Vergne. IV, 240 pages. 1977.

Vol. 588: P. Molino, Théorie des G-Structures: Le Problème d'Equivalence. VI, 163 pages. 1977.

Vol. 589: Cohomologie l-adique et Fonctions L. Séminaire de Géométrie Algébrique du Bois-Marie 1965–66, SGA 5. Edité par L. Illusie. XII, 484 pages. 1977.

Vol. 590: H. Matsumoto, Analyse Harmonique dans les Systèmes de Tits Bornologiques de Type Affine. IV, 219 pages. 1977.

Vol. 591: G. A. Anderson, Surgery with Coefficients. VIII, 157 pages. 1977.

Vol. 592: D. Voigt, Induzierte Darstellungen in der Theorie der endlichen, algebraischen Gruppen. V, 413 Seiten. 1977.

Vol. 593: K. Barbey and H. König, Abstract Analytic Function Theory and Hardy Algebras. VIII, 260 pages. 1977.

Vol 594: Singular Perturbations and Boundary Layer Theory, Lyon 1976. Edited by C. M Brauner, B. Gay, and J. Mathieu. VIII, 539 pages. 1977.

Vol. 595: W. Hazod, Stetige Faltungshalbgruppen von Wahrscheinlichkeitsmaßen und erzeugende Distributionen. XIII, 157 Seiten. 1977.

Vol. 596: K. Deimling, Ordinary Differential Equations in Banach Spaces. VI, 137 pages. 1977

Vol. 597: Geometry and Topology, Rio de Janeiro, July 1976. Proceedings. Edited by J. Palis and M. do Carmo. VI, 866 pages. 1977.

Vol. 598: J. Hoffmann-Jørgensen, T. M. Liggett et J. Neveu, Ecole d'Eté de Probabilités de Saint-Flour VI – 1976. Edité par P.-L. Hennequin. XII, 447 pages. 1977.

Vol. 599: Complex Analysis, Kentucky 1976. Proceedings. Edited by J. D. Buckholtz and T. J. Suffridge. X, 159 pages. 1977.

Vol. 600: W. Stoll, Value Distribution on Parabolic Spaces. VIII, 216 pages. 1977.

Vol. 601: Modular Functions of one Variable V, Bonn 1976. Proceedings Edited by J.-P. Serre and D. B. Zagier. VI, 294 pages. 1977.

Vol. 602: J. P. Brezin, Harmonic Analysis on Compact Solvmanifolds. VIII, 179 pages. 1977.

Vol. 603: B. Moishezon, Complex Surfaces and Connected Sums of Complex Projective Planes. IV, 234 pages. 1977

Vol. 604: Banach Spaces of Analytic Functions, Kent, Ohio 1976. Proceedings. Edited by J. Baker, C. Cleaver and Joseph Diestel. VI, 141 pages. 1977.

Vol. 605: Sario et al., Classification Theory of Riemannian Manifolds. XX, 498 pages. 1977.

Vol. 606: Mathematical Aspects of Finite Element Methods. Proceedings 1975. Edited by I. Galligani and E. Magenes. VI, 362 pages. 1977.

Vol. 607: M. Métivier, Reelle und Vektorwertige Quasimartingale und die Theorie der Stochastischen Integration. X, 310 Seiten. 1977.

Vol. 608: Bigard et al., Groupes et Anneaux Réticulés. XIV, 334 pages. 1977.